Lecture Notes in Computer Science   10932

Commenced Publication in 1973
Founding and Former Series Editors:
Gerhard Goos, Juris Hartmanis, and Jan van Leeuwen

More information about this series at http://www.springer.com/series/7407

Yutaka Takahashi · Tuan Phung-Duc
Sabine Wittevrongel · Wuyi Yue (Eds.)

Queueing Theory and Network Applications

13th International Conference, QTNA 2018
Tsukuba, Japan, July 25–27, 2018
Proceedings

 Springer

Editors
Yutaka Takahashi
Kyoto University
Kyoto
Japan

Sabine Wittevrongel ⓘ
Ghent University
Gent
Belgium

Tuan Phung-Duc
University of Tsukuba
Tsukuba
Japan

Wuyi Yue
Konan University
Kobe
Japan

ISSN 0302-9743   ISSN 1611-3349   (electronic)
Lecture Notes in Computer Science
ISBN 978-3-319-93735-9   ISBN 978-3-319-93736-6   (eBook)
https://doi.org/10.1007/978-3-319-93736-6

Library of Congress Control Number: 2018947320

LNCS Sublibrary: SL1 – Theoretical Computer Science and General Issues

Printed on acid-free paper

This Springer imprint is published by the registered company Springer International Publishing AG
part of Springer Nature
The registered company address is: Gewerbestrasse 11, 6330 Cham, Switzerland

Preface

The International Conference on Queueing Theory and Network Applications aims to promote the knowledge and the development of high-quality research on queueing theory and its applications in networks and other related fields. It brings together researchers, scientists, and practitioners from over the world and offers an open forum to share the latest important research accomplishments and challenging problems in the area of queueing theory and network applications.

This volume contains papers selected and presented at the 13th International Conference on Queueing Theory and Network Applications (QTNA 2018) held during July 25–27, 2018, in Tsukuba, Ibaraki, Japan.

QTNA 2018 was a continuation of the series of successful QTNA conferences: QTNA 2006 (Seoul, Korea), QTNA 2007 (Kobe, Japan), QTNA 2008 (Taipei, Taiwan), QTNA 2009 (Singapore), QTNA 2010 (Beijing, China), QTNA 2011 (Seoul, Korea), QTNA 2012 (Kyoto, Japan), QTNA 2013 (Taichung, Taiwan), QTNA 2014 (Bellingham, USA), QTNA 2015 (Hanoi, Vietnam), QTNA 2016 (Wellington, New Zealand), and QTNA 2017 (Qinhuangdao, China).

The conference this year was truly international, having received 57 submissions from 23 countries and areas in five continents: Algeria, Australia, Austria, Belgium, Bulgaria, Canada, China, Colombia, Hong Kong, Hungary, Iceland, India, Israel, Italy, Japan, The Netherlands, Russia, South Korea, Sri Lanka, Taiwan, Turkey, USA, and Vietnam. These papers were peer reviewed and evaluated on the quality, originality, soundness, and significance of their contributions by the members of the Technical Program Committee (TPC) of QTNA 2018 and external reviewers invited by the TPC. Each paper was reviewed by at least three reviewers. After a careful selection, eight full papers (12+ pages) and ten short papers (6–11 pages) were accepted for inclusion in this volume of *Lecture Notes in Computer Science* (LNCS) published by Springer.

Furthermore, 20 papers (2–5 pages) showing on-going research were selected for presentation at the conference and for inclusion in an electronic version of the conference brochure that was distributed to all the participants of QTNA 2018.

All the papers to be presented disseminate the latest results covering up-to-date research fields such as performance modeling and analysis of telecommunication systems, retrial and vacation queueing models, optimization of queueing systems, modeling of social systems, and application of machine learning in queueing models.

It was our privilege to invite Professor Benny Van Houdt to give a keynote talk and Professor Moshe Haviv to deliver a tutorial lecture at QTNA 2018.

We would like to thank the authors of all the papers appearing in these proceedings for their excellent contribution. Special thanks go to the co-chairs and members of the Technical Program Committee of QTNA 2018 for their time and effort to assuring the quality of the selected papers. We also would like to express our gratitude to the co-chairs and members of the local Organizing Committee for their hard work throughout the process from planning to holding the conference. Finally, we cordially thank Springer for their support in publishing this volume.

July 2018

Yutaka Takahashi
Tuan Phung-Duc
Sabine Wittevrongel
Wuyi Yue

Organization

Honorary Chair

Hisashi Kobayashi   Princeton University, USA

Organizing Committee

General Chairs

Hideaki Takagi   University of Tsukuba, Japan
Yutaka Takahashi   Kyoto University, Japan

Members

Shoji Kasahara, Japan
Quan-lin Li, China
Hsing Paul Luh, Taiwan
Tuan Phung-Duc, Japan
Sabine Wittevrongel, Belgium
Wuyi Yue, Japan

Steering Committee

Co-chairs

Bong Dae Choi   Korea University, Korea
Yutaka Takahashi   Kyoto University, Japan
Wuyi Yue   Konan University, Japan

Members

Hsing Paul Luh, Taiwan
Winston K. G. Seah, New Zealand
Hideaki Takagi, Japan
Y. C. Tay, Singapore
Kuo-Hsiung Wang, Taiwan
Jinting Wang, China
Dequan Yue, China
Zhe George Zhang, USA

Technical Program Committee

Co-chairs

Tuan Phung-Duc   University of Tsukuba, Japan
Sabine Wittevrongel   Ghent University, Belgium
Wuyi Yue   Konan University, Japan

Members

Herwig Bruneel, Belgium
Wai-Ki Ching, Hong Kong
Wanyang Dai, China
Tien V. Do, Hungary
Alexander Dudin, Belarus
Antonis Economou, Greece
Pengfei Guo, Hong Kong
Guangyue Han, Hong Kong
Qi-Ming He, Canada
Ganguk Hwang, Korea
Yoshiaki Inoue, Japan
Shunfu Jin, China
Ken'ichi Kawanishi, Japan
Konosuke Kawashima, Japan
Jau-Chuan Ke, Taiwan
Jyh-Bin Ke, Taiwan
Bara Kim, Korea
Masahiro Kobayashi, Japan
Achyutha Krishnamoorthy, India
Ho Woo Lee, Korea
Se Won Lee, Korea
Tony T. Lee, Hong Kong
Bin Liu, China
Zhanyou Ma, China
Hiroyuki Masuyama, Japan
Agassi Melikov, Azerbaijan
Zhisheng Niu, China
Rein D. Nobel, The Netherlands
Toshihisa Ozawa, Japan
Wouter Rogiest, Belgium
Poompat Saengudomlert, Thailand
Zsolt Saffer, Hungary
Yutaka Sakuma, Japan
Winston K. G. Seah, New Zealand
Sergey Andreev, Finland
Yang Woo Shin, Korea
Zhankun Sun, Hong Kong

Ahmed Mohamed Kamel Tarabia, Egypt
Y. C. Tay, Singapore
Miklos Telek, Hungary
Koen De Turck, France
Jinting Wang, China
Kuo-Hsiung Wang, Taiwan
Hengqing Ye, Hong Kong
Xue-Ming Yuan, Singapore
Dequan Yue, China
Zhe George Zhang, USA
Yigiang Q. Zhao, Canada

Local Organizing Committee

Co-chairs

Wuyi Yue	Konan University, Japan
Tuan Phung-Duc	University of Tsukuba, Japan

Members

Atsushi Inoie	Kanagawa Institute of Technology, Japan
Yoshiaki Inoue	Osaka University, Japan
Shoji Kasahara	Nara Institute of Science and Technology, Japan
Ken'ichi Kawanishi	Gunma University, Japan
Tatsuaki Kimura	NTT Network Technology Laboratories
Masahiro Kobayashi	Tokai University, Japan
Hiroyuki Masuyama	Kyoto University, Japan
Yutaka Sakuma	National Defense Academy, Japan
Yuanyuan Wang	Yamaguchi University, Japan
Moeko Yajima	Tokyo Institute of Technology, Japan

Contents

Queueing Models with Retrials

A Priority Retrial Queue with Constant Retrial Policy

Arnaud Devos$^{(\boxtimes)}$, Joris Walraevens, and Herwig Bruneel

SMACS Research Group, Department of Telecommunications and Information
Processing (EA07), Ghent University - UGent,
Sint-Pietersnieuwstraat 41, 9000 Gent, Belgium
{arnaud.devos,joris.walraevens,herwig.bruneel}@ugent.be

Abstract. We analyse a priority queueing system with a normal queue
(high priority) and an orbit (low priority). Only the first customer in orbit
can retry during times that the queue and server are empty (constant
retrial policy). In contrast with existing literature, we assume different
service time distributions for the high- and low-priority customers. We
obtain closed-form expressions for the probability generating function of
the number of customers in queue and orbit, in steady state, and for the
Laplace Stieltjes transforms of the stationary waiting times of both type
of customers.

Keywords: Priority retrial queues · Constant retrial policy

1 Introduction

In some queueing systems, customers who find the server busy upon arrival can
decide to come back to the system after some random time. Those customers
can be visualised as being located in a virtual (retrial) queue, hereafter called
the orbit. In queueing theory, these queues are labelled as retrial queues. In
the literature, two types of retrial policies are typically identified. The first one,
called the classical retrial policy, assumes that all the customers in the orbit
retry to get access to the server. For analytical feasibility reasons, these retrials
are typically modelled as independent Poisson processes (with the same rate
for each customer). Typical examples in such a situation are call centres and
restaurant reservations. An excellent survey of these models can be found in [2].
The second policy, referred to as the constant retrial policy, assumes that only
the head of the orbit can retry for service [5]. This policy was first introduced
in [6], where a telephone exchange model with only one line is examined. In
[6], blocked customers (incoming calls) form a feedback queue (this is the orbit)
and only the customer at the head of the feedback queue can retry for service
after an exponentially distributed retrial time. A more recent application uses
a constant retrial queue to model the behaviour of call blending in call centers
[9,10]. In contrast to the classical retrial policy, the retrial times in the constant

retrial policy are independent of the orbit length. Therefore, the distribution of the retrial times does not have to be restricted to an exponential one to obtain closed-form expressions for performance measures. In [8], a retrial queue with constant retrial policy and general retrial times is extensively studied. The author has obtained the stationary distribution for the orbit length, the server state and the waiting time.

If retrial queues also possess a regular queue, priority policy occurs naturally. Customers in the regular queue have priority over the customers in the orbit. This means that the customer at the head of the orbit can only start retrying when the regular queue becomes empty. If another customer arrives during a retrial time, this customer is served and the retrial has to start over when the regular queue becomes empty again. This priority can be non-preemptive or preemptive. In [7], a preemptive constant retrial policy is studied. The author has obtained the stationary distribution for the queue and orbit length, and the waiting time of both type of customers.

In [1], a non-preemptive constant retrial policy is studied. The stationary distributions for the queue and orbit lengths are obtained. An interesting result from this paper, is that the distribution of the priority queue is independent of the retrial time distribution. While in [7] it is assumed that the service time distributions of the customers in the priority queue and customers in the orbit are different, this is not the case in [1]. However, there might be situations where the service time distributions are not necessary the same. It might even be the reason why a customer joins either the queue or the orbit. For instance, customers with a small (non-urgent) service time who face a large queue upon arrival, are (in certain situations) more likely to join the orbit. Introducing a regular queue also causes that the distribution of the waiting time of customers in the orbit is harder to analyse than, for example, in [8]. Besides the difficulty of the retrial policy, an additional difficulty arises due to the priority mechanism. Arrivals in the regular queue during service times have to be served before the waiting customers in the orbit.

In this paper, we generalize the model in [1] in two ways: we assume that both type of customers have different service time distributions. Further, in contrast with [1], we also analyse the waiting times of both type of customers. The remainder of this paper is outlined as follows. In the next section we provide a more detailed description of the queueing model under consideration. In Sect. 3 we analyse the steady state distribution of the queue and orbit lengths, using the supplementary variable technique [3]. Section 4 gives a brief overview of the most important performances measures deduced from the steady state distributions. An extensive analysis of the waiting times is done in Sect. 5. Finally, we discuss some numerical examples in Sect. 6.

2   Mathematical Model

We assume a queueing system with an infinite-sized queue, an infinite-sized orbit, and one server. Class-1 and class-2 customers arrive to the system according to two independent Poisson processes with rate λ_1 and λ_2 respectively. If a new customer arrives to the system and the server is free, he gets service immediately. If the server is busy upon arrival, a class-1 customer waits in the queue, while a class-2 customer joins the orbit. We assume that only the first customer in the orbit retries to get access to the server when the queue and server are idle. Retrial times are characterized by means of i.i.d. continuous random variables with common cumulative distribution (cdf) $R(x)$ and common Laplace Stieltjes transform (LST) $R^*(s)$. Service times of class-j customers, $j = 1, 2$, are specified by i.i.d. continuous random variables with cdf B_j, LST $B_j^*(s)$ and mean β_j. We define the arrival load as $\rho \triangleq \lambda_1\beta_1 + \lambda_2\beta_2$. Further, we assume that service times, retrial times and inter-arrival times are all mutually independent.

In the rest of this section we will fix notations for the rest of the paper.

We use the notation $\mathbb{P}$ for the probability measure, $\mathbb{E}$ for the expectation operator and $\mathbb{1}$ for the indicator function.

Let $N_1(t)$ be the number of class-1 customers in the priority queue at time t, not counting the one in the server. Similarly, let $N_2(t)$ the number of class-2 customers in the orbit at time t. Further, we add four auxiliary processes: $C(t) = 0$ when the server is idle at time t, $C(t) = 1$ when the server is serving a class-1 customer at time t and $C(t) = 2$ when the server is serving a class-2 customer at time t. When $C(t) = 0$ and $N_2(t) > 0$, $\xi_0(t)$ represents the elapsed retrial time at time t; when $C(t) = 1$, $\xi_1(t)$ denotes the elapsed class-1 service time at time t; when $C(t) = 2$, $\xi_2(t)$ denotes the elapsed class-2 service time at time t. The stochastic vector $\{C(t), N_1(t), N_2(t), \xi_0(t), \xi_1(t), \xi_2(t)\}$ is then a Markov Process. We assume that the stationary distribution of this stochastic vector exists.[1] Therefore we define the following limiting probabilities:

$$p_0 = \lim_{t\to\infty} \mathbb{P}[C(t) = N_1(t) = N_2(t) = 0],$$

$$p_{0,n}^{(0)}(x)dx = \lim_{t\to\infty} \mathbb{P}[C(t) = 0, N_1(t) = 0, N_2(t) = n, x < \xi_0(t) \le x + dx], \ n \ge 1$$

$$p_{i,n}^{(1)}(x)dx = \lim_{t\to\infty} \mathbb{P}[C(t) = 1, N_1(t) = i, N_2(t) = n, x < \xi_1(t) \le x + dx], \ i, n \ge 0$$

$$p_{i,n}^{(2)}(x)dx = \lim_{t\to\infty} \mathbb{P}[C(t) = 2, N_1(t) = i, N_2(t) = n, x < \xi_2(t) \le x + dx], \ i, n \ge 0.$$

We also define the following probability generating functions (pgf)

$$P_0(x, z_2) = \sum_{n=1}^{\infty} p_{0,n}^{(0)}(x) \, z_2^n, \tag{1}$$

$$P_1(x, z_1, z_2) = \sum_{i=0}^{\infty} \sum_{n=0}^{\infty} p_{i,n}^{(1)}(x) \, z_1^i z_2^n, \tag{2}$$

---

[1] A sufficient condition for this assumption could be derived by first studying the embedded Markov chain at departure epochs, similar as in [1]. It can be shown that this condition turns out to be the same as the necessary stability condition we will encounter in Sect. 3.

$$P_2(x, z_1, z_2) = \sum_{i=0}^{\infty} \sum_{n=0}^{\infty} p_{i,n}^{(2)}(x)\, z_1^i z_2^n. \tag{3}$$

Next, we define $\bar{r}(x)$ as the probability density function (pdf) of the retrial time R, given that the retrial time is at least x. We define $\bar{b}_j(x)$ as the pdf of the service time of a class-j customer, given that the service time is at least x. We thus have

$$\bar{r}(x) = \frac{R'(x)}{1 - R(x)} \quad \text{and} \quad \bar{b}_j(x) = \frac{B_j'(x)}{1 - B_j(x)}. \tag{4}$$

Let $\lambda \triangleq \lambda_1 + \lambda_2$ the total arrival rate to the system. Let p_1 and p_2 the probability that a customer is of class-1 and of class-2 respectively. These probabilities are given by $p_j = \lambda_j/\lambda$, $j = 1, 2$.

For further use, we define $A_j(z_1, z_2)$ as the joint pgf of the class-1 and class-2 arrivals during a class-j service time and $A(z_1, z_2)$ as the joint pgf of the class-1 and class-2 arrivals during a randomly chosen service time. They are calculated as

$$A_j(z_1, z_2) = B_j^*(\lambda_1(1 - z_1) + \lambda_2(1 - z_2)), \tag{5}$$
$$A(z_1, z_2) = p_1 A_1(z_1, z_2) + p_2 A_2(z_1, z_2). \tag{6}$$

In particular, we are interested in the steady-state distributions of the orbit and queue length. We define

$$Q(z) = p_0 + \int_0^{\infty} P_0(x, z)dx, \tag{7}$$

as the partial pgf of the number of customers in the orbit when the server is idle, and

$$U_j(z_1, z_2) = \int_0^{\infty} P_j(x, z_1, z_2)dx, \quad j = 1, 2, \tag{8}$$

as the partial joint pgf of the number of customers in queue and orbit when the server is serving a type-j customer.

3   Queue and Orbit Lengths

If the regular queue is non-empty, a customer from that queue is served next. Otherwise, a customer from the orbit (if any) retries to access the server. If his retrial time finishes before a new arrival, he is served. Otherwise, the newly arriving customer starts service, irrespective of his type, and the customer at the head of the orbit stops retrying until the server and the priority queue are empty again. This leads to the following equilibrium equations

$$\lambda p_0 = \int_0^\infty p_{0,0}^{(1)}(x)\bar{b}_1(x)dx + \int_0^\infty p_{0,0}^{(2)}(x)\bar{b}_2(x)dx, \tag{9}$$

$$\frac{\partial}{\partial x}p_{0,n}^{(0)}(x) = -(\lambda + \bar{r}(x))p_{0,n}^{(0)}(x), \quad n \geq 1, \tag{10}$$

$$\frac{\partial}{\partial x}p_{i,n}^{(1)}(x) = -(\lambda + \bar{b}_1(x))p_{i,n}^{(1)}(x) + \lambda_1 p_{i-1,n}^{(1)}(x)\mathbb{1}(i \geq 1)$$
$$+ \lambda_2 p_{i,n-1}^{(1)}(x)\mathbb{1}(n \geq 1), \quad i \geq 0, n \geq 0, \tag{11}$$

$$\frac{\partial}{\partial x}p_{i,n}^{(2)}(x) = -(\lambda + \bar{b}_2(x))p_{i,n}^{(2)}(x) + \lambda_1 p_{i-1,n}^{(2)}(x)\mathbb{1}(i \geq 1)$$
$$+ \lambda_2 p_{i,n-1}^{(2)}(x)\mathbb{1}(n \geq 1), \quad i \geq 0, n \geq 0. \tag{12}$$

The boundary conditions are

$$p_{0,n}^{(0)}(0) = \int_0^\infty p_{0,n}^{(1)}(x)\bar{b}_1(x)dx + \int_0^\infty p_{0,n}^{(2)}(x)\bar{b}_2(x)dx, \quad n \geq 1, \tag{13}$$

$$p_{i,n}^{(1)}(0) = \mathbb{1}(i = 0, n = 0)\lambda_1 p_0 + \mathbb{1}(i = 0, n > 0)\lambda_1 \int_0^\infty p_{0,n}^{(0)}(x)dx$$
$$+ \int_0^\infty p_{i+1,n}^{(1)}(x)\bar{b}_1(x)dx + \int_0^\infty p_{i+1,n}^{(2)}(x)\bar{b}_2(x)dx, \quad i \geq 0, n \geq 0, \tag{14}$$

$$p_{i,n}^{(2)}(0) = \mathbb{1}(i = 0, n = 0)\lambda_2 p_0 + \mathbb{1}(i = 0, n > 0)\lambda_2 \int_0^\infty p_{0,n}^{(0)}(x)dx$$
$$+ \mathbb{1}(i = 0)\int_0^\infty p_{0,n+1}^{(0)}(x)\bar{r}(x)dx, \quad i \geq 0, n \geq 0. \tag{15}$$

Assuming stationarity, the equilibrium equations (10), (11) and (12) are transformed to the following partial differential equations for the pgfs $P_0(x,z)$, $P_1(x,z_1,z_2)$ and $P_2(x,z_1,z_2)$:

$$\frac{\partial}{\partial x}P_0(x,z_2) = -(\lambda + \bar{r}(x))P_0(x,z_2),$$

$$\frac{\partial}{\partial x}P_1(x,z_1,z_2) = -(\lambda + \bar{b}_1(x) - \lambda_1 z_1 - \lambda_2 z_2)P_1(x,z_1,z_2),$$

$$\frac{\partial}{\partial x}P_2(x,z_1,z_2) = -(\lambda + \bar{b}_2(x) - \lambda_1 z_1 - \lambda_2 z_2)P_1(x,z_1,z_2),$$

whose solutions are given by

$$P_0(x,z_2) = P_0(0,z_2)[1 - R(x)]e^{-\lambda x}, \tag{16}$$

$$P_1(x,z_1,z_2) = P_1(0,z_1,z_2)[1 - B_1(x)]e^{-(\lambda_1(1-z_1)+\lambda_2(1-z_2))x}, \tag{17}$$

$$P_2(x,z_1,z_2) = P_2(0,z_1,z_2)[1 - B_2(x)]e^{-(\lambda_1(1-z_1)+\lambda_2(1-z_2))x}. \tag{18}$$

In terms of the pgfs, the boundary conditions can be written as

$$P_0(0, z_2) = \int_0^\infty P_1(x, 0, z_2)\bar{b}_1(x)dx + \int_0^\infty P_2(x, 0, z_2)\bar{b}_2(x)dx - \lambda p_0, \quad (19)$$

$$P_1(0, z_1, z_2) = \lambda_1 p_0 + \lambda_1 \int_0^\infty P_0(x, z_2)dx$$

$$+ \frac{1}{z_1}\int_0^\infty [P_1(x, z_1, z_2) - P_1(x, 0, z_2)]\bar{b}_1(x)dx$$

$$+ \frac{1}{z_1}\int_0^\infty [P_2(x, z_1, z_2) - P_2(x, 0, z_2)]\bar{b}_2(x)dx, \quad (20)$$

$$P_2(0, z_1, z_2) = \lambda_2 p_0 + \lambda_2 \int_0^\infty P_0(x, z_2)dx + \frac{1}{z_2}\int_0^\infty P_0(x, z_2)\bar{r}(x)dx, \quad (21)$$

where we also used (9) in the first equation.

If we substitute (16), (17) and (18) in the previous equations and by using (5), we get

$$P_0(0, z_2) = P_1(0, 0, z_2)A_1(0, z_2) + P_2(0, 0, z_2)A_2(0, z_2) - \lambda p_0, \quad (22)$$

$$P_1(0, z_1, z_2) = \lambda_1 p_0 + p_1(1 - R^*(\lambda))P_0(0, z_2) + \frac{1}{z_1}P_1(0, z_1, z_2)A_1(z_1, z_2)$$

$$- \frac{1}{z_1}P_1(0, 0, z_2)A_1(0, z_2) + \frac{1}{z_1}P_2(0, z_1, z_2)A_2(z_1, z_2)$$

$$- \frac{1}{z_1}P_2(0, 0, z_2)A_2(0, z_2), \quad (23)$$

$$P_2(0, z_1, z_2) = \lambda_2 p_0 + \frac{1}{z_2}[p_2 z_2(1 - R^*(\lambda)) + R^*(\lambda)]P_0(0, z_2). \quad (24)$$

We are left with five unknowns: p_0, $P_0(0, z_2)$, $P_1(0, z_1, z_2)$, $P_1(0, 0, z_2)$ and $P_2(0, 0, z_2)$.[2] First, we substitute (24) in (22) and solve for $P_0(0, z_2)$,

$$P_0(0, z_2) = \frac{z_2 A_1(0, z_2)P_1(0, 0, z_2) + (p_2 A_2(0, z_2) - 1)z_2\lambda p_0}{z_2 - [p_2 z_2 + R^*(\lambda)(1 - p_2 z_2)]A_2(0, z_2)}. \quad (25)$$

Substituting (24) in (23) and solving to $P_1(0, z_1, z_2)$, we can write

$$z_2[z_1 - A_1(z_1, z_2)]P_1(0, z_1, z_2) = \{z_1 p_1 + p_2 A_2(z_1, z_2) - p_2 A_2(0, z_2)\}z_2\lambda p_0$$
$$+ \{z_2(1 - R^*(\lambda))[p_1 z_1 + p_2 A_2(z_1, z_2) - p_2 A_2(0, z_2)]$$
$$+ R^*(\lambda)[A_2(z_1, z_2) - A_2(0, z_2)]\} P_0(0, z_2) - z_2 A_1(0, z_2)P_1(0, 0, z_2). \quad (26)$$

It is well-known that $f(z_1) := z_1 - A_1(z_1, z_2)$ has one zero inside the complex unit disk for each z_2 with $|z_2| < 1$, see for instance [12]. Call this zero $Y(z_2)$. It can be proven that this is an analytic function inside the complex unit disk and that $Y(1) = 1$, at least when the queueing system is stable. For more details, we

---

[2] Note from Eq. (24) that $P_2(0, z_1, z_2)$ is independent of z_1.

refer to Appendix A in [13]. By substituting z_1 by $Y(z_2)$ in (26), we find

$$
\begin{aligned}
z_2 A_1(0, z_2) P_1(0, 0, z_2) = {} & \{A(Y(z_2), z_2) - p_2 A_2(0, z_2)\} z_2 \lambda p_0 \\
& + \{z_2(1 - R^*(\lambda))[A(Y(z_2), z_2) - p_2 A_2(0, z_2)] \\
& + R^*(\lambda)[A_2(Y(z_2), z_2) - A_2(0, z_2)]\} P_0(0, z_2).
\end{aligned} \tag{27}
$$

Substituting the above equation into (25) yields

$$
P_0(0, z_2) = \frac{[A(Y(z_2), z_2) - 1] z_2 \lambda p_0}{z_2 - z_2(1 - R^*(\lambda)) A(Y(z_2), z_2) - R^*(\lambda) A_2(Y(z_2), z_2)}. \tag{28}
$$

Substituting Eq. (28) back into (25) and solving for $P_1(0, 0, z_2)$ yields

$$
\begin{aligned}
P_1(0, 0, z_2) = {} & \frac{R^*(\lambda) \lambda p_0}{z_2 - z_2(1 - R^*(\lambda)) A(Y(z_2), z_2) - R^*(\lambda) A_2(Y(z_2), z_2)} \\
& \times \frac{[1 - A(Y(z_2), z_2))][A_2(0, z_2) - z_2] + [p_2 A_2(0, z_2) - 1][A_2(Y(z_2), z_2) - z_2]}{A_1(0, z_2)}.
\end{aligned} \tag{29}
$$

Inserting (28) and (29) into (26) gives

$$
\begin{aligned}
P_1(0, z_1, z_2) = {} & \frac{R^*(\lambda) \lambda p_0}{z_2 - z_2(1 - R^*(\lambda)) A(Y(z_2), z_2) - R^*(\lambda) A_2(Y(z_2), z_2)} \\
& \times \left\{ \frac{[1 - p_1 z_1 - p_2 A_2(z_1, z_2)][A_2(Y(z_2), z_2) - z_2]}{z_1 - A_1(z_1, z_2)} \right. \\
& \left. + \frac{[A_2(z_1, z_2) - z_2][A(Y(z_2), z_2) - 1]}{z_1 - A_1(z_1, z_2)} \right\}.
\end{aligned} \tag{30}
$$

The forth unknown $P_2(0, z_1, z_2)$ is given as

$$
P_2(0, z_1, z_2) = \frac{[p_2 z_2 + p_1 Y(z_2) - 1] R^*(\lambda) \lambda p_0}{z_2 - z_2(1 - R^*(\lambda)) A(Y(z_2), z_2) - R^*(\lambda) A_2(Y(z_2), z_2)}. \tag{31}
$$

Finally, the last unknown p_0 can be found from the normalisation condition $Q(1) + U_1(1, 1) + U_2(1, 1) = 1$. Using the definitions (7), (8) and the expressions (16), (17) and (18), one easily finds

$$
Q(z_2) = p_0 + \frac{1 - R^*(\lambda)}{\lambda} P_0(0, z_2),
$$

$$
U_j(z_1, z_2) = \frac{1 - A_j(z_1, z_2)}{\lambda_1(1 - z_1) + \lambda_2(1 - z_2)} P_j(0, z_1, z_2), \quad j = 1, 2.
$$

It now follows that

$$
p_0 = 1 - \frac{R^*(\lambda) p_1 + p_2}{R^*(\lambda)} \rho. \tag{32}
$$

Requiring $p_0 > 0$ leads to a necessary stability condition for the system.

Summarized, we have obtained partial pgfs of the number of customers in the queue and orbit when the server is either busy or idle. The complete expressions are given by:

$$Q(z) = \frac{z - A_2(Y(z), z)}{z - z(1 - R^*(\lambda))A(Y(z), z) - R^*(\lambda)A_2(Y(z), z)}$$
$$\times \left(1 - \frac{R^*(\lambda)p_1 + p_2}{R^*(\lambda)}\rho\right) R^*(\lambda), \tag{33}$$

$$U_1(z_1, z_2) = \left\{ \frac{[1 - p_1 z_1 - p_2 A_2(z_1, z_2)][A_2(Y(z_2), z_2) - z_2]}{z_2 - z_2(1 - R^*(\lambda))A(Y(z_2), z_2) - R^*(\lambda)A_2(Y(z_2), z_2)} \right.$$
$$+ \left. \frac{[A_2(z_1, z_2) - z_2][A(Y(z_2), z_2) - 1]}{z_2 - z_2(1 - R^*(\lambda))A(Y(z_2), z_2) - R^*(\lambda)A_2(Y(z_2), z_2)} \right\}$$
$$\times \frac{1 - A_1(z_1, z_2)}{z_1 - A_1(z_1, z_2)} \frac{1}{1 - p_1 z_1 - p_2 z_2} \left(1 - \frac{R^*(\lambda)p_1 + p_2}{R^*(\lambda)}\rho\right) R^*(\lambda), \tag{34}$$

$$U_2(z_1, z_2) = \frac{p_2 z_2 + p_1 Y(z_2) - 1}{z_2 - z_2(1 - R^*(\lambda))A(Y(z_2), z_2) - R^*(\lambda)A_2(Y(z_2), z_2)}$$
$$\times \frac{1 - A_2(z_1, z_2)}{1 - p_1 z_1 - p_2 z_2} \left(1 - \frac{R^*(\lambda)p_1 + p_2}{R^*(\lambda)}\rho\right) R^*(\lambda). \tag{35}$$

From these three important pgfs, the marginal pgfs of the numbers of customers in the queue and orbit are easily deduced. The pgf of N_1 is given by

$$Q(1) + U_1(z, 1) + U_2(z, 1) = 1 - \rho + \frac{1 - p_1 \rho - A(z, 1) + p_1 \rho A_1(z, 1)}{p_1[A_1(z, 1) - z]}. \tag{36}$$

The pgf of N_1 is thus independent of the numbers of customers in the orbit ánd the distribution of the retrial times. As mentioned before, this was also observed in [1]. The pgf of N_2 is given by

$$Q(z) + U_1(1, z) + U_2(1, z) = \frac{p_1 Y(z) + p_2 z - 1}{z - z(1 - R^*(\lambda))A(Y(z), z) - R^*(\lambda)A_2(Y(z), z)}$$
$$\times \frac{1}{p_2} \left(1 - \frac{R^*(\lambda)p_1 + p_2}{R^*(\lambda)}\rho\right) R^*(\lambda). \tag{37}$$

4   Performance Measures

In this section we list some important performance measures of the queueing system.

The probability that the server is empty is given by $Q(1)$,

$$Q(1) = 1 - \rho, \tag{38}$$

It is worth noting that this is not equal to the probability that the system is empty, which is given by p_0, see (32).

The probability that the server is occupied by a class-j customer is given by $U_j(1,1)$,

$$U_j(1,1) = \lambda_j b_j, \quad j = 1, 2. \tag{39}$$

The mean length of the priority queue is found by taking the first derivative of (36), yielding

$$\mathbb{E}[N_1] = \frac{\lambda_1(\lambda_1 B_1^{*''}(0) + \lambda_2 B_2^{*''}(0))}{2(1 - \beta_1 \lambda_1)}. \tag{40}$$

The mean orbit length is found by taking the first derivative of (37), which leads to

$$\mathbb{E}[N_2] = \frac{\lambda_2[p_2 + p_1 R^*(\lambda)][\lambda_1 B_1^{*''}(0) + \lambda_2 B_2^{*''}(0)]}{2(1 - \lambda_1\beta_1)(R^*(\lambda) - [p_2 + R^*(\lambda)p_1]\rho)} \\ + \frac{\lambda_2[1 - R^*(\lambda)](p_1\beta_1 + p_2\beta_2)}{R^*(\lambda) - [p_2 + R^*(\lambda)p_1]\rho}. \tag{41}$$

Another interesting performance measure is the mean size of the orbit when the server is idle. This measure is given by

$$Q'(1) = (1 - R^*(\lambda))p_2 \\ \times \frac{\lambda_2(\lambda_1 B_1^{*''}(0) + \lambda_2 B_2^{*''}(0)) + 2(1 - \lambda_1\beta_1)\rho(1 - \rho)}{2(1 - \lambda_1\beta_1)(R^*(\lambda) - [p_2 + R^*(\lambda)p_1]\rho)}. \tag{42}$$

5 Waiting Time

In this section we analyse the waiting times of class-1 and class-2 customers in steady state. The waiting time of a customer starts when the customer enters the system and ends when the customer enters the server. The waiting time of a generic class-j customer is denoted by W_j, $j = 1, 2$, and its LST by $W_j^*(s)$. Tag a randomly arriving class-j customer and denote him by C_j. The distribution of the queue length and the orbit length at the arrival instant of C_j, is the same as at an arbitrary time point because of the PASTA property.

5.1 Class-1 Customers

When C_1 arrives to the system, there are three possibilities: (1) the server is idle, (2) the server is serving a class-1 customer or (3) the server is serving a class-2 customer.

When the server is idle, C_1 gets service immediately (the priority queue must be idle in this case) and consequently has no waiting time. Hence,

$$\mathbb{E}[e^{-sW_1}\mathbb{1}\{\text{no service}\}] = 1 - \rho. \tag{43}$$

When the server is busy, the waiting time of C_1 consists of the remaining service time of the customer currently in service and the service times of all class-1 customers that are in the priority queue. This yields

$$W_1 = \sum_{k=1}^{N_1} B_{1,k} + T_j, \quad \text{if the server is serving a class-}j\text{-customer}, \tag{44}$$

where T_j, $j = 1, 2$, denotes the remaining service time and $B_{1,k}$ denotes the service time of the k-th customer in the queue. The $B_{1,k}$ are i.i.d. with LST $B_1^*(s)$ and are independent of T_j. The distribution of the remaining service time depends on the class of the customer being served and the elapsed service time. Its conditional LST is given by

$$\mathbb{E}[e^{-sT_j}|C = j, x < \xi_j \le x + dx] = \frac{1}{1 - B_j(x)} \int_x^\infty e^{-s(y-x)} dB_j(y), \quad j = 1, 2.$$
(45)

Hence, using (44), it follows that for $j = 1, 2$, $i, n \in \mathbb{N}$ and $x > 0$:

$$\mathbb{E}[e^{-sW_1}|C = j, N_1 = i, N_2 = n, x < \xi_j \le x + dx]$$
$$= B_1^*(s)^i \left(\frac{1}{1 - B_j(x)} \int_x^\infty e^{-s(y-x)} dB_j(y) \right). (46)$$

Averaging over all possible i, n and x, we find

$$\mathbb{E}[e^{-sW_1} \mathbb{1}\{\text{service class-}j\}]$$

$$= \int_0^\infty \sum_{i=0}^\infty \sum_{n=0}^\infty B_1^*(s)^i \left(\frac{1}{1 - B_j(x)} \int_x^\infty e^{-s(y-x)} dB_j(y) \right) p_{i,n}^{(j)}(x) dx$$

$$= \int_0^\infty \left(\frac{1}{1 - B_j(x)} \int_x^\infty e^{-s(y-x)} dB_j(y) \right) P_j(x, B_1^*(s), 1) dx$$

$$= P_j(0, B_1^*(s), 1) \int_0^\infty \int_x^\infty e^{-s(y-x)} e^{-\lambda_1(1 - B_1^*(s))x} dB_j(y) dx$$

$$= P_j(0, B_1^*(s), 1) \int_0^\infty e^{-sy} \int_0^y e^{(s - \lambda_1(1 - B_1^*(s)))x} dx \, dB_j(y)$$

$$= \frac{P_j(0, B_1^*(s), 1)[A_j(B_1^*(s), 1) - B_j^*(s)]}{s - \lambda_1(1 - B_1^*(s))}.$$
(47)

The full expressions can be found by using (30):

$$\mathbb{E}[e^{-sW_1} \mathbb{1}\{\text{service class-}1\}] = \frac{\lambda_1(B_1^*(s) - 1)(1 - \rho) + \lambda_2[A_2(B_1^*(s), 1) - 1]}{B_1^*(s) - A_1(B_1^*(s), 1)}$$
$$\times \frac{A_1(B_1^*(s), 1) - B_1^*(s)}{s - \lambda_1(1 - B_1^*(s))},$$
(48)

and by using (31):

$$\mathbb{E}[e^{-sW_1} \mathbb{1}\{\text{service class-}2\}] = \lambda_2 \frac{A_2(B_1^*(s), 1) - B_2^*(s)}{s - \lambda_1(1 - B_1^*(s))}.$$
(49)

We can now calculate the LST of W_1:

$$
\begin{aligned}
W_1^*(s) &\triangleq \mathbb{E}[e^{-sW_1}] \\
&= \mathbb{E}[e^{-sW_1}\mathbb{1}\{\text{no service}\}] + \mathbb{E}[e^{-sW_1}\mathbb{1}\{\text{service class-1}\}] \\
&\quad + \mathbb{E}[e^{-sW_1}\mathbb{1}\{\text{service class-2}\}] \\
&= 1 - \rho \\
&\quad + \frac{\lambda_1(B_1^*(s)-1)(1-\rho) + \lambda_2[A_2(B_1^*(s),1)-1]}{B_1^*(s) - A_1(B_1^*(s),1)} \cdot \frac{A_1(B_1^*(s),1) - B_1^*(s)}{s - \lambda_1(1 - B_1^*(s))} \\
&\quad + \lambda_2 \frac{A_2(B_1^*(s),1) - B_2^*(s)}{s - \lambda_1(1 - B_1^*(s))}.
\end{aligned}
\tag{50}
$$

One can verify that

$$
\mathbb{E}[W_1] = \frac{\lambda_1 B_1^{*\prime\prime}(0) + \lambda_2 B_2^{*\prime\prime}(0)}{2(1 - \lambda_1\beta_1)},
\tag{51}
$$

which is in accordance with Little's Law.

5.2 Class-2 Customers

Because of the priority mechanism, finding an expression for $W_2^*(s)$ is more involved. Class-1 and class-2 arrivals during retrial times have to be served before C_2. Moreover, class-1 arrivals during service times while C_2 is in the orbit have to be served before C_2 as well. This difficulty leads to the concept of "sub-busy periods" [11]. We will define two kinds of sub-busy periods, i.e., sub-busy periods initiated by class-1 customers and sub-busy periods initiated by class-2 customers.

The sub-busy period initiated by a class-1 customer starts at the time instant he enters the server and ends when the number of class-1 customers in the system is decreased by 1 for the first time.

The second type - a sub-busy period initiated by a class-2 packet - starts at the time instant the initiating class-2 customers enters the server and ends at the beginning of which a new class-2 customer can start to retrial (if any).

Denote the LST of a type-j sub-busy period as $V_j^*(s)$, $j = 1,2$. The LST can be expressed as the following functional equations:

$$
V_j^*(s) = B_j^*(s + \lambda_1(1 - V_1^*(s))), \quad j = 1, 2.
\tag{52}
$$

This can be understood as follows. Denote by V_j a type-j busy period and B_j the service time of the customer initiating the sub-busy period. Suppose that when the service starts, there are m customers in the queue. During B_j, a number of class-1 customers may arrive, say $\tilde{a}_1$. Hence, the number of customers in the queue at the end of the service is equal to $m + \tilde{a}_1 - 1$. The sub-busy period ends when there are for the first time $m - 1$ customers in the queue. To achieve this, one needs $\tilde{a}_1$ additional sub-busy periods. One can thus write

$$
V_j = B_j + \sum_{k=1}^{\tilde{a}_1} V_{1,k}.
\tag{53}
$$

The $V_{1,k}$ are i.i.d. and have LST $V_1^*(s)$. Equation (52) then follows by taking the LST of (53), taking into account that $\tilde{a}_1$ is the number of arrivals during B_j.

We distinguish again three possibilities when C_2 arrives to the system: (1) the server is free, (2) the server is occupied by a class-1 customer or (3) the server is occupied by a class-2 customer.

When the server is free, C_2 gets immediate access to the server and has no waiting time. Hence,

$$\mathbb{E}[e^{-sW_2}\mathbb{1}\{\text{no service}\}] = 1 - \rho. \tag{54}$$

When the server is busy, the waiting time of C_2 consists of the remaining service time of the customer currently in service, the sub-busy periods initiated by the class-1 arrival during that remaining service time, the sub-busy period initiated by the class-1 customers in the queue, the effective retrial times[3] of the class-2 customers in the orbit, the sub-busy periods initiated by the class-2 customers in the orbit, and the effective retrial time of C_2. This yields,

$$W_2 = T_j + \sum_{k=1}^{a^{(T_j)}} V_{1,k} + \sum_{k=1}^{N_1} V_{1,k} + \sum_{k=1}^{N_2}(V_{2,k} + G_k) + G, \tag{55}$$

where T_j, $j = 1, 2$, denotes the remaining service time and $a^{(T_j)}$ the number of class-1 arrivals during T_j. $V_{j,k}$ denotes the k-th sub-busy period initiated by the k-th class-j customer and G_k the effective retrial time of the k-th class-2 customer in the orbit. Finally, G denotes the effective retrial time of C_2.

It is clear that the lengths of the sub-busy periods initiated by class-j customers are i.i.d. and thus all have the same LST $V_j^*(s)$, $j = 1, 2$. Because the retrial times are assumed i.i.d., the same holds for the effective retrial times, thus they all have the same LST $G^*(s)$.

Apart from the first two terms in (55), all random variables are independent, for given values of N_1 and N_2. It holds, for $j = 1, 2$,

$$\mathbb{E}\left[e^{-s\left(T_j + \sum_{k=1}^{a^{(T_j)}} V_{1,k}\right)} \middle| C = j, x < \xi_j \leq x + dx\right]$$

$$= \frac{1}{1 - B_j(x)} \int_x^\infty e^{-(s+\lambda_1(1-V_1^*(s)))(y-x)} dB_j(y). \tag{56}$$

We now concentrate on the effective retrial time of a random class-2 customer. The effective retrial time of a class-2 customer equals the initial retrial time if no arrivals occur during his retrial time or equals the time of an unsuccessful attempt plus a sub busy period and the remaining effective retrial time after this attempt, if this is not the case. That is,

$$G = \begin{cases} S & \text{no arrivals during retrial time,} \\ F + V_1 + \tilde{G} & \text{retrial interrupted by a class-1 arrival,} \\ F + V_2 + \tilde{G} & \text{retrial interrupted by a class-2 arrival.} \end{cases} \tag{57}$$

---

[3] The effective retrial time of class-2 customer starts when the customer is for the first time the head of the orbit and ends when he enters the server.

Here G and $\tilde{G}$ denote the effective retrial time and the remaining effective retrial time after an unsuccessful attempt. Because the retrial time is resampled after an unsuccessful attempt, G and $\tilde{G}$ have the same distribution. S denotes the successful retrial time and F denotes the time of an unsuccessful retrial. The distribution of F is independent of the type of the arriving customers that interrupts it, because the minimum of two exponentially distributed random variables is independent of which one of the two is the minimum [4, Chap. 2]. The reason why we make a distinction in the type of customer that interrupted the retrial time, is because the initiated sub-busy period distribution depends on this type.

Due to the competition of two exponentials with rate λ_j, $j = 1, 2$ and a general retrial time, it follows that

$$\mathbb{P}[\text{no arrivals during retrial}] = R^*(\lambda),$$
$$\mathbb{P}[\text{retrial interrupted by a class-1 arrival}] = p_1(1 - R^*(\lambda)),$$
$$\mathbb{P}[\text{retrial interrupted by a class-2 arrival}] = p_2(1 - R^*(\lambda)).$$

Let T be an exponential with rate λ and R a general retrial time. The cdf of a successful retrial time can be found as

$$\mathbb{P}[R \leq t | R < T] = \frac{\int_0^t e^{-\lambda y} dR(y)}{R^*(\lambda)}.$$

The LST of this distribution is then given as

$$S^*(s) \triangleq \mathbb{E}[e^{-sR} | R < T] = \frac{R^*(\lambda + s)}{R^*(\lambda)}. \tag{58}$$

The cdf of an interrupted retrial time is

$$\mathbb{P}[T \leq t | T < R] = \frac{\int_0^t \lambda e^{-\lambda y}(1 - R(y)) dy}{(1 - R^*(\lambda))},$$

which leads to

$$F^*(s) \triangleq \mathbb{E}[e^{-sT} | T < R] = \frac{\lambda(1 - R^*(\lambda + s))}{(s + \lambda)(1 - R^*(\lambda))}. \tag{59}$$

Taking the LST of (57), using the previous expressions and solving for $G^*(s)$ yields

$$G^*(s) = \frac{(s + \lambda)R^*(\lambda + s)}{s + \lambda - [1 - R^*(\lambda + s)][\lambda_1 V_1^*(s) + \lambda_2 V_2^*(s)]}. \tag{60}$$

So far, we have obtained every distribution which occurs in (55). It holds for $j = 1, 2$, $i, n \in \mathbb{N}$, $x > 0$:

$$\mathbb{E}[e^{-sW_2} | C = j, N_1 = i, N_2 = n, x < \xi_j \leq x + dx]$$
$$= \left(\frac{1}{1 - B_j(x)} \int_x^\infty e^{-(s + \lambda_1(1 - V_1^*(s)))(y - x)} dB_j(y) \right)$$
$$\times V_1^*(s)^i [G^*(s) V_2^*(s)]^n G^*(s). \tag{61}$$

Averaging over all possible i, n and x, we find, in a similar way as (47),

$$\mathbb{E}[e^{-sW_2}\mathbb{1}\{\text{service class-}j\}] = \frac{G^*(s)P_j\left(0, V_1^*(s), V_2^*(s)G^*(s)\right)}{s - \lambda_2(1 - V_2^*(s)G^*(s))}$$
$$\times \left[A_j(V_1^*(s), V_2^*(s)G^*(s)) - V_j^*(s)\right]. \quad (62)$$

We can now calculate the LST of W_2,

$$W_2^*(s) \triangleq \mathbb{E}[e^{-sW_2}]$$
$$= \mathbb{E}[e^{-sW_2}\mathbb{1}\{\text{no-service}\}] + \mathbb{E}[e^{-sW_2}\mathbb{1}\{\text{service class-1}\}]$$
$$+ \mathbb{E}[e^{-sW_2}\mathbb{1}\{\text{service class-2}\}]$$
$$= 1 - \rho$$
$$+ \frac{G^*(s)P_1\left(0, V_1^*(s), V_2^*(s)G^*(s)\right)}{s - \lambda_2(1 - V_2^*(s)G^*(s))}\left[A_1(V_1^*(s), V_2^*(s)G^*(s)) - V_1^*(s)\right]$$
$$+ \frac{G^*(s)P_2\left(0, V_1^*(s), V_2^*(s)G^*(s)\right)}{s - \lambda_2(1 - V_2^*(s)G^*(s))}\left[A_2(V_1^*(s), V_2^*(s)G^*(s)) - V_2^*(s)\right]. \quad (63)$$

One can again verify Little's Law:

$$\mathbb{E}[W_2] = \frac{[p_2 + p_1 R^*(\lambda)][\lambda_1 B_1^{*''}(0) + \lambda_2 B_2^{*''}(0)]}{2(1 - \lambda_1\beta_1)(R^*(\lambda) - [p_2 + R^*(\lambda)p_1]\rho)}$$
$$+ \frac{[1 - R^*(\lambda)](p_1\beta_1 + p_2\beta_2)}{R^*(\lambda) - [p_2 + R^*(\lambda)p_1]\rho}. \quad (64)$$

6   Numerical Examples

To conclude this paper, we discuss some numerical examples. We will specifically focus on the influence of the service times and the retrial policy.

We assume the service times of type-1 customers to be exponentially distributed with mean β_1, i.e.

$$B_1^*(s) = (1 + \beta_1 s)^{-1}. \quad (65)$$

In order to study the influence of different service times, we assume the service times of type-2 customers to be of the gamma type with shape parameter $1/2$ and mean β_2, i.e.

$$B_2^*(s) = (1 + 2\beta_2 s)^{-1/2}. \quad (66)$$

The reason for this distribution is twofold. The first reason is to have two different service time distributions. The second reason is to have a coefficient of variation[4] that is bigger than one, which is equal to $\sqrt{2}$. The coefficient of variation for $B_1^*(s)$ obviously equals one.

---

[4] The coefficient of variation of a distribution is defined as the ratio of the standard deviation to the mean.

Next, we suppose the retrial times follow an Erlang distribution of order two and rate $\nu > 0$, i.e.

$$R^*(s) = \left(\frac{\nu}{\nu + s}\right)^2.$$ (67)

Finally, we define α as the fraction of class-1 customers load in the overall traffic mix, i.e.,

$$\alpha = \frac{\lambda_1 \beta_1}{\rho},$$ (68)

with $\rho = \lambda_1 \beta_1 + \lambda_2 \beta_2$ as before.

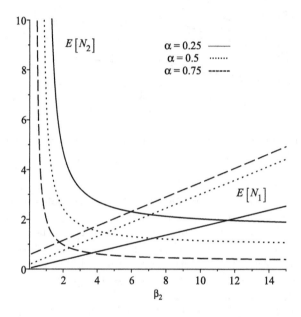

Fig. 1. Mean values of the queue and orbit lengths versus the mean class-2 service times ($\rho = 0.7$, $\beta_1 = 1$ and $\nu = 3$).

Figures 1 and 2 show the mean queue length and mean orbit length, and the mean waiting times respectively versus the mean service time of class-2 customers with $\rho = 0.7$, $\beta_1 = 1$, $\nu = 3$ and $\alpha = 0.25$, 0.5 and 0.75. From Fig. 1, it can be seen that the mean queue length increases linearly, while the mean orbit length decreases with increasing β_2. The decrease of the mean orbit length is caused by the decrease of λ_2 when β_2 increases (ρ is kept constant), hence less class-2 customers arrive to the system. The increase in the mean queue length can be explained as follows. When β_2 increases, more class-1 customers will arrive during the time periods that a class-2 customer is being served. Due to the non-preemptive priority policy, arriving class-1 customers cannot interrupt the service of class-2 customers. Hence, they have to wait until this class-2 customer leaves the system, which results in a larger mean queue length.

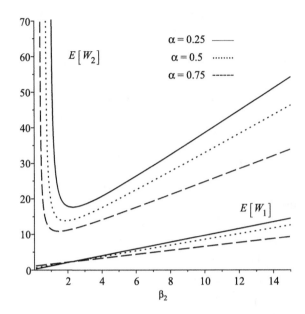

Fig. 2. Mean values of the waiting times versus the mean class-2 service times ($\rho = 0.7$, $\beta_1 = 1$ and $\nu = 3$).

Looking at Fig. 2, we observe that the (mean) waiting time of class-2 customers first decreases and then increases, for increasing β_2. This effect is due to the retrial policy. To understand this, we present in Fig. 3 the same model as in Fig. 2, but where the retrial times are zero, i.e. $R^*(s) = 1$. In this way, we obtain the ordinary non-preemptive priority queueing model because class-2 customers do not have to retrial any more when the server gets empty. Obviously, the mean waiting times for class-1 customers are the same in both figures, because the waiting time distribution of class-1 is independent of the retrial distribution. The decrease for small β_2 observed in Fig. 2 can be explained as follows. For very small β_2, λ_2 is large and hence the orbit size is high (as observed in Fig. 1). All these class-2 customers have a small service time, but their retrial time is not necessary small. This actually depends on the retrial rate ν. Also, the probability of a successful retrial, i.e. $R^*(\lambda_1 + \lambda_2)$, decreases as λ_2 increases (for fixed λ_1). This can be seen by taking the first derivative of (67). Consequently, new class-2 arrivals will face a large waiting time when β_2 is very small. If β_2 increases, λ_2 decreases, which results in a decrease for the waiting time for class-2 customers because less customers have to retrial. This *positive* effect continues until β_2 achieves a critical value and from there, the negative effect of longer service times dominates the positive effect of less customers.

We now repeat the same experiment, with varying mean service time of class-1 customers while keeping the mean service time of class-2 customers fixed. We set $\rho = 0.65$, $\nu = 3$, $\beta_2 = 1$ and $\alpha = 0.25$, 0.5 and 0.75. Figure 4 shows the mean queue length and mean orbit length versus the mean service time of

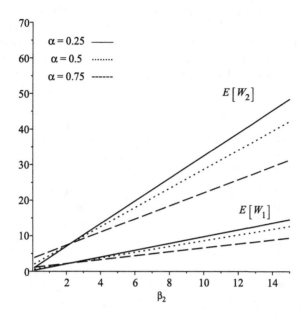

Fig. 3. Mean values of the waiting times versus the mean class-2 service times ($\rho = 0.7$, $\beta_1 = 1$ and $R^*(s) = 1$).

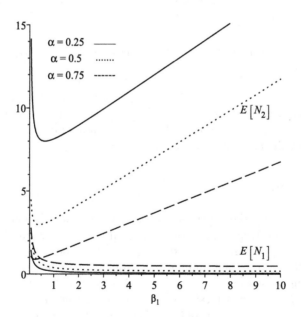

Fig. 4. Mean values of the queue and orbit lengths versus the mean class-1 service times ($\rho = 0.65$, $\beta_2 = 1$ and $\nu = 3$).

class-1 customers. The reason why the mean queue length slightly decreases as β_1 increases is due to decreasing λ_1 for increasing β_1. Apart from very small values of β_1, the mean orbit size strongly increases as β_1 because the server will be less available. The opposite effect for (very) small values can again be explained by the previous discussion.

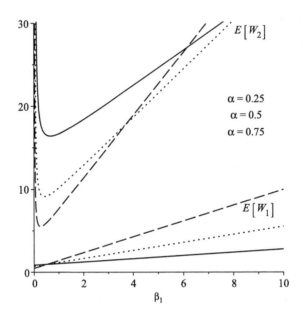

Fig. 5. Mean values of the waiting times versus the mean class-1 service times ($\rho = 0.65$, $\beta_2 = 1$ and $\nu = 3$.)

The influence of the mean service time of class-1 customers on the waiting times is shown in Fig. 5. One can observe that the minimum in the upper curves is much sharper than in the previous example.

7   Conclusion

In this paper, we analysed a priority retrial queue for a constant retrial policy and different service time distributions. The supplementary variable technique was used to obtain partial pgfs for the number of customers in the orbit when the server is idle and the joint number of customers in queue and orbit when the server is serving a type-1 customer, respectively a type-2 customer. The LST of the stationary waiting times of both type of customers is obtained. Using numerical examples, we illustrated the influence of the service times and the retrial policy on the mean queue length, the mean orbit length and the mean waiting times respectively.

An interesting future research direction is the analysis of the asymptotic behaviour of type-2 customers and compare these results with the asymptotics for the corresponding model in the classical retrial setting.

Acknowledgments. The authors wish to thank Tuan Phung-Duc and Dieter Claeys for preliminary discussions about the model in this paper.

References

1. Atencia, I., Moreno, P.: A single-server retrial queue with general retrial times and Bernoulli schedule. Appl. Math. Comput. **162**(2), 855–880 (2005)
2. Choi, B.D., Chang, Y.: Single server retrial queues with priority calls. Math. Comput. Modell. **30**, 7–32 (1999)
3. Cox, D. : The analysis of non-Markovian stochastic processes by the inclusion of supplementary variables. In: Mathematical Proceedings of the Cambridge Philosophical Society, vol. 51, no. 3, pp. 433–441. Cambridge University Press (1955)
4. Durrett, R.: Essentials of Stochastic Processes. STS. Springer, New York (2012). https://doi.org/10.1007/978-1-4614-3615-7
5. Farahmand, K.: Single line queue with repeated demands. Queueing Syst. **6**(1), 223–228 (1990)
6. Fayolle, G.: A simple telephone exchange with delayed feedbacks. In: Proceedings of the International Seminar on Teletraffic Analysis and Computer Performance Evaluation, pp. 245–253. North-Holland Publishing Co. (1986)
7. Gao, S.: A preemptive priority retrial queue with two classes of customers and general retrial times. Oper. Res. Int. J. **15**(2), 233–251 (2015)
8. Gómez-Corral, A.: Stochastic analysis of a single server retrial queue with general retrial times. Naval Res. Logistics (NRL) **46**, 561–581 (1999)
9. Phung-Duc, T., Rogiest, W.: Two way communication retrial queues with balanced call blending. In: Al-Begain, K., Fiems, D., Vincent, J.-M. (eds.) ASMTA 2012. LNCS, vol. 7314, pp. 16–31. Springer, Heidelberg (2012). https://doi.org/10.1007/978-3-642-30782-9_2
10. Saffer, Z., Yue, W.: Call blending retrial queue with gated type incoming call service. In: 9th International Conference on Queueing Theory and Network Applications, Bellingham, USA, pp. 18–27 (2014)
11. Walraevens, J., Steyaert, B., Bruneel, H.: Delay characteristics in discrete-time GI-G-1 queues with non-preemptive priority queueing discipline. Perform. Eval. **50**(1), 53–75 (2002)
12. Walraevens, J., Steyaert, B., Bruneel, H.: Performance analysis of a single-server ATM queue with a priority scheduling. Comput. Oper. Res. **30**(12), 1807–1829 (2003)
13. Walraevens, J.: Discrete-time queueing models with priorities. Doctoral dissertation, Ghent University (2004)

Regenerative Analysis of Two-Way Communication Orbit-Queue with General Service Time

Evsey Morozov[1,2] and Tuan Phung-Duc[3(✉)]

[1] Institute of Applied Mathematical Research, Karelian Research Centre of RAS,
11 Pushkinskaya Street, 185910 Petrozavodsk, Russia
[2] Petrozavodsk State University, Petrozavodsk, Russia
emorozov@karelia.ru
[3] Faculty of Engineering, Information and Systems, University of Tsukuba, Tsukuba,
Ibaraki, Japan
tuan@sk.tsukuba.ac.jp

Abstract. In this paper, we apply a regenerative approach to reprove some recent steady-state results [1,8,9] for an orbit-queue (also known as retrial queue with a constant retrial rate) with outgoing calls. Stability conditions are discussed as well. Moreover, some generalizations of the model are also considered.

Keywords: Stability condition · Retrial system · Classical discipline
Outgoing calls · Regenerative approach · General service time

1 Introduction

1.1 Motivation and Related Work

In call centers with the callback option, customers (incoming calls) who cannot connect immediately with the operator, register their number to be called back in a later time [2]. Since these customers are not present at the call center, they cannot be picked up immediately when the operator is available. Instead, even when the operator becomes available, some seeking time is needed to access a registered customer. In what follows, we assume that registered customers stay in an orbit-queue. The seeking process can be done either manually by the operator or automatically by a machine. Furthermore, apart from incoming calls, the operator may make outgoing calls in its idle time. An outgoing call may be referred to as a private job of the operator or a real call to a customer outside. It should be noted that in case the seeking mechanism is carried out by a machine, the operator may not be aware of the presence of registered customers in the orbit-queue. Thus, the operator may make an outgoing call even when there are some registered customers in the orbit-queue. Furthermore, even in the case the operator is aware of registered customers in the orbit-queue, he may also make outgoing calls if the outgoing calls are more urgent and/or important than

© Springer International Publishing AG, part of Springer Nature 2018
Y. Takahashi et al. (Eds.): QTNA 2018, LNCS 10932, pp. 22–32, 2018.
https://doi.org/10.1007/978-3-319-93736-6_2

incoming calls. The same phenomenon is also observed in various service systems where a ticket is issued upon the arrival of a customer who cannot be served immediately. The customer will be called in a later time when the server is available.

Motivated by this real situation, a single server queueing system with an orbit-queue has been proposed and analyzed in [8,9] under pure Markovian settings, i.e., the arrival process of incoming calls is Poisson and the service time distributions of incoming and outgoing calls are exponential. The orbit-queue (also called the orbit in the literature) corresponds to the queue for registered customers in call centers. Recently, Aissani and Phung-Duc [1] have analyzed the single server model with arbitrarily distributed incoming calls and outgoing calls and Poisson input of incoming calls. The analysis of [1] is based on the balance equations for the underlying Markov process of the system. They have obtained explicit formulae for the generating functions of the joint queue-length distribution and exact formulae for the mean orbit sizes.

1.2   Contribution of the Current Paper

In this paper, using a much simpler alternative approach, we obtain some explicit performance measures such as the distribution of the state of the server, the probability of an empty system etc. for the model in [1]. Furthermore, the method is applied to obtain some performance measures for the multiple server model and the single server model with the general renewal input.

More precisely, the contribution of this work is two-fold. First of all, we apply the regenerative methodology to reprove some important performance results established for this model in previous works [1,8,9]. We emphasize that the new proofs in general are much simpler. However, not all the results obtained in [1,8, 9] can be regenerated by the analysis of this paper. It is the feature of the applied methodology: it does not use detailed balance equations describing the dynamics of the process, unlike the approach based on the Kolmogorov equations for the corresponding Markov process. On the other hand, the regenerative approach is not limited by the framework of Markov processes, and allows to analyze much wider class of queueing processes though in less detail. We emphasize that the full power of the regenerative methodology is demonstrated by stability analysis, but it will not be elaborate in this paper.

1.3   Organization of the Paper

The paper is organized as follows. The model is described in Sect. 2, where also basic notation is given. In Sect. 3, we first prove main results for single server system with one class of the outgoing calls. Then, we consider the multiclass system, and finally, the system with m parallel servers. In Sect. 4, we discuss sufficient stability conditions. In particular, we give stability condition for the model with two classes of outgoing calls with general service times.

2   Description of the Model

We consider a single-server retrial queue with two-way communication. Primary incoming calls (customers) arrive at the server (or operator) according to a Poisson process with instants $\{t_n,\ n \geq 1\}$ with rate $\lambda \in (0,\ \infty)$. Incoming calls finding an idle server receive service immediately. The service times of the incoming calls $\{S_n,\ n \geq 1\}$ are independent identically distributed (iid) with a general distribution with mean $\mathsf{E}S =: 1/\mu < \infty$ (Here and in what follows, we omit the serial index to denote a generic element of an iid sequence).

In case of a busy server, the incoming call enters an orbit. Within the orbit, a constant retrial policy is applied, i.e., the arrival rate of calls from the orbit is μ_0 (if the orbit is not-empty) and independent of the number customers being in the orbit. In other words, retrial attempts follow exponential distribution with rate μ_0 (we denote this random variable $\exp(\mu_0)$). A constant retrial rate occurs when customers form a FCFS queue in the orbit, and only the customer at the head of the queue can request service. Another option is that the system has a control mechanism allowing to keep a constant summary retrial rate regardless of the number of orbital calls, although these calls may make the retrial attempts independently.

We consider the case where the server makes outgoing calls in its idle time. More precisely, when the server is idle for an exponentially distributed time with rate γ_1, it makes an outgoing call. It is equivalent to that outgoing calls arrive at the server according to a Poisson process with rate γ_1 and if the server is busy upon arrival of an outgoing call, the outgoing call is lost. As a result, after the completion of each service, the next action of the server is determined by the competition of three events, i.e., an arrival of a fresh incoming call, a retrial of an incoming call from the orbit queue (if the orbit is not empty) and an arrival of an outgoing call.

The service times of outgoing calls $\{S_n^{(1)},\ n \geq 1\}$ are iid with a general distribution with mean $\mathsf{E}S^{(1)} =: 1/\mu_1 < \infty$.

For simplicity, we assume zero initial state: the 1st customer arrives in the idle system at instant $t = 0$. In interval $[0,\ t)$, we denote $V(t)$ the sum of the service times of all incoming customers arriving in the system and all outgoing calls, $B(t)$ the busy time of the server, $I(t)$ the idle time of the server, that is $B(t) + I(t) = t$. Denote $N(t)$ the number of orbital customers and $W(t)$ the workload (remaining work) in orbit queue at instant t^-. Let $S(t)$ be the remaining service time at instant t^-, i.e. the time the current customer, being in server, departs server. (By definition, $S(t) = 0$ if $Q(t) = 0$). We consider the following basic one-dimensional *non-Markovian* process $X(t) := N(t) + Q(t),\ t \geq 0$, where $Q(t) \in \{0,\ 1\}$ is the number of calls in the server at instant t^-. We emphasize that a reduction of the dimension of a basic process is a crucial advantage of the regenerative approach. Denote $X(t_n) = X_n,\ n \geq 1$. Put $T_0 = 0$ and define recursively,

$$T_{n+1} = \inf\left(t_k > T_n : X_k = 0\right),\ n \geq 0.$$

It is easy to see that $\{T_n\}$ are classical regenerations of the basic process X, with the iid *regeneration periods* $T_{n+1} - T_n$ (and with generic period T). It is well-known (for instance, [3]) that if the mean generic period $\mathsf{E}T < \infty$ then the process $\{X(t)\}$ (and the basic system) is *positive recurrent* and (because the input flow is Poisson) the steady-state distribution of the process $\{X(t)\}$ (and other related regenerative processes) exists. In other words, there exists the weak limit $X(t) \Rightarrow X$. (The critical role of the requirement $\mathsf{E}T < \infty$ for stability analysis is discussed in detail in [4]). We first assume positive recurrent case, $\mathsf{E}T < \infty$, and later on discuss in brief stability conditions which imply this basic requirement.

Denote the following stationary probabilities: the server is idle P_0; the server is busy $\mathsf{P}_b = 1 - \mathsf{P}_0$; the server is busy by an incoming call $\mathsf{P}_b^{(0)}$; the server is busy by an outgoing call $\mathsf{P}_b^{(1)}$; the system (orbit and server) is empty, $\mathsf{P}_{0,0}$; and finally, the stationary probability that server is empty and orbit is busy, $\mathsf{P}_{0,b}$. As we mentioned above, positive recurrence implies the existence of these stationary probabilities. For instance, $\lim_{t\to\infty} \mathsf{P}(X(t) = 0) = \mathsf{P}_{0,0}$, $\lim_{t\to\infty} \mathsf{P}(Q(t) = 0, N(t) > 0) = \mathsf{P}_{0,b}$. Also denote

$$\rho = \frac{\lambda}{\mu}, \ \sigma = \frac{\gamma_1}{\mu_1}.$$

Below we obtain, by a simple regenerative approach, some important steady-state performance measures of the model which have been found in [8,9] by the detailed Kolmogorov equations approach, for pure Markovian setting.

3 Stability Analysis

3.1 One Class of Outgoing Calls

Using the regenerative approach, we below prove the following results which have been proved in [8] for pure Markovian model by Kolmogorov equations approach.

Theorem 1. The stationary probabilities satisfy the following relations:

$$\mathsf{P}_0 = \frac{1-\rho}{1+\sigma} = 1 - \mathsf{P}_b; \tag{1}$$

$$\mathsf{P}_b^{(0)} = \rho; \tag{2}$$

$$\mathsf{P}_b^{(1)} = \sigma \frac{1-\rho}{1+\sigma}; \tag{3}$$

$$\mathsf{P}_b = \mathsf{P}_b^{(0)} + \mathsf{P}_b^{(1)} = \frac{\sigma+\rho}{1+\sigma}; \tag{4}$$

$$\mathsf{P}_{0,0} = \frac{1 - \lambda/\mu_0[\rho + \sigma + \mu_0/\mu]}{1+\sigma}; \tag{5}$$

$$\mathsf{P}_{0,b} = \sigma \frac{\lambda}{\mu_0} \frac{(1-\rho)}{(1+\sigma)}. \tag{6}$$

Proof. Denote $A_0(t)$ the number of incoming arrivals in $[0, t)$, and let $A_1(I(t))$ be the number of the outgoing calls which appear during (summary) idle period $I(t)$ of server in period $[0, t)$. Also denote $W_0(t)$ the remaining work in orbit at instant t. We note that the amount of work $V(t)$ generated (in interval $[0, t)$) by the outgoing calls and incoming customers, equals the sum of the remaining work $(W_0(t) + S(t))$ and the work which departs system in interval $[0, t)$. The latter in turns equals the busy time $B(t)$ in interval $[0, t)$. Thus we obtain the following balance equation:

$$V(t) = \sum_{i=1}^{A_0(t)} S_i + \sum_{i=1}^{A_1(I(t))} S_i^{(1)}$$
$$= S(t) + W_0(t) + t - I(t). \tag{7}$$

We note that the remaining service time $S(t)$ may relate both to the incoming call and outgoing call, however in either case $S(t) = o(t)$ (see [5]) and, because of the positive recurrence, $W_0(t) = o(t)$, $t \to \infty$ with probability 1 (w. p. 1) as well (see [6] and [10]). Moreover, by the Strong Law of Large Numbers (SLLN) for the renewal process, w.p.1,

$$\lim_{t \to \infty} \frac{A_0(t)}{t} = \lambda, \ \lim_{t \to \infty} \frac{1}{A_0(t)} \sum_{i=1}^{A_0(t)} S_i = \frac{1}{\mu}, \ \lim_{t \to \infty} \frac{A_1(t)}{t} = \gamma_1. \tag{8}$$

In addition, by the positive recurrence, the following limits exist

$$\lim_{t \to \infty} \frac{I(t)}{t} = \mathsf{P}_0 = \lim_{t \to \infty} \mathsf{P}(Q(t) = 0) = \mathsf{P}(Q = 0), \tag{9}$$

where Q denotes the stationary number of calls in the server. Note that the first limit in (9) holds w.p.1 because $I(t)$ is a cumulative process with positive recurrent embedded process of regenerations [10].

Then, by (8) and (9), we obtain w.p.1,

$$\frac{V(t)}{t} = \frac{A_0(t)}{t} \frac{1}{A_0(t)} \sum_{i=1}^{A_0(t)} S_i + \frac{A_1(I(t))}{I(t)} \frac{I(t)}{t} \frac{\sum_{i=1}^{A_1(I(t))} S_i^{(1)}}{A_1(I(t))}$$
$$\to \rho + \sigma \mathsf{P}_0, \ t \to \infty. \tag{10}$$

On the other hand,

$$\lim_{t \to \infty} \frac{1}{t} \Big(S(t) + W_0(t) + t - I(t) \Big) = 1 - \mathsf{P}_0. \tag{11}$$

Now (1) follows from (7)–(11). To establish (2), we write down the following balance equation for the work $V_0(t)$ which incoming customers bring in the system in interval $[0, t)$:

$$V_0(t) = S_0(t) + B_0(t) + W_0(t), \tag{12}$$

where $S_0(t)$ is the remaining service time of an incoming call, provided it is in the server ($S_0(t) = 0$, if the server is free or serves an outgoing call), and $B_0(t)$ is the busy time the server devotes to incoming calls in interval $[0, t)$. Note that, as above, $S_0(t) = W_0(t) = o(t)$ w.p.1. Because $V_0(t) = \sum_{i=1}^{A_0(t)} S_i$, then (2) immediately follows from (12) and the following relations hold w.p.1:

$$\frac{V_0(t)}{t} \to \rho, \quad \frac{B_0(t)}{t} \to \mathsf{P}_b^{(0)}, \quad t \to \infty.$$

To find $\mathsf{P}_{0,b}$, we denote $D_0(t)$ the number of retrial calls which would depart the orbit in $[0, t)$ *provided the server devotes all its time to these calls*. Note that the actual number of the incoming calls which join the orbit in interval $[0, t)$ stochastically equals $A_0(B(t))$. (Indeed we use here the property of the Poisson input). Then we have the following balance relation between the number of incoming calls joint to orbit and the number of retrial calls departed orbit, in interval $[0, t)$:

$$A_0(B(t)) = N(t) + D_0(L(t)), \tag{13}$$

where

$$L(t) := \{t : N(t) > 0, \, Q(t) = 0\}$$

is the summary time when the orbit is busy while server is free, allowing successful retrial attempts. Note that this representation is again correct due to Poisson inputs formed by the incoming calls and the retrial calls attempting to enter server. By the stationarity, $N(t) = o(t)$, and we have

$$\lim_{t \to \infty} \frac{1}{t} A_0(B(t)) = \lambda \mathsf{P}_b; \tag{14}$$

$$\lim_{t \to \infty} \frac{1}{t}(N(t) + D_0(L(t))) = \mu_0 \mathsf{P}_{0,b}, \tag{15}$$

where we take into account that, by the SLLN for the renewal process D_0,

$$\frac{D_0(L(t))}{L(t)} \to \mu_0,$$

and that there exists the limit

$$\lim_{t \to \infty} \frac{L(t)}{t} = \mathsf{P}_{0,b}.$$

Finally, we write

$$\mathsf{P}_{0,0} = \mathsf{P}_0 - \mathsf{P}_{0,b} = \frac{1}{1+\sigma}[1 - \rho - \frac{\lambda}{\mu_0}(\sigma + \rho)], \tag{16}$$

which is equivalent to (5). ∎

Now, we consider the stability condition.

Let N_n be the orbit size *just after the nth departure* from the system, then the server is free at this instant, and we denote it $Q_n = 0$. It is easy to see that $\{N_n\}$ is the Markov chain. Assumed positive recurrence of the basic process implies also the existence of the stationary distribution of this chain. In particular, there exists the stationary probability (in evident notation)

$$\lim_{n\to\infty} \mathsf{P}(N_n = 0, Q_n = 0) = \mathsf{P}_{0,0}.$$

Now it is clear that the necessary stability condition of the original model is $\mathsf{P}_{0,0} > 0$.

Thus we obtain the following necessary stability condition of the model under consideration.

Corollary. If the original model with general service times (both for incoming and outgoing calls) is stable (positive recurrent) then

$$\rho + \frac{\lambda}{\mu_0}(\sigma + \rho) < 1. \tag{17}$$

It is easy to see that it is the same stability condition which has been proved in [1,8].

We recall the definition of the classical retrial policy [7]. In this case the retrial rate depends on the orbit size. In the simplest case, the rate is proportional to the orbit size. In other words, if μ_0 is individual retrial rate of each call and orbit size equals n then summary retrial rate equals $\mu_0 n$. Also we note that stability criterion of this model is $\rho < 1$ and does not include the retrial rate, see for instant, [7]. A careful analysis of the above given proof allows to conclude that some of the obtained results stay valid also for this retrial model with classical retrial policy. Namely, the following statement holds.

Theorem 2. The statements (1)–(4) stay valid for the retrial system with classical retrial policy.

3.2   M Classes of Outgoing Calls

The analysis of the stationary regime developed above can be easily extended to the same single-server system with Poisson incoming calls with rate λ and with an arbitrary number M of classes of the outgoing calls, denoted by $1,\dots,M$. Denote $\{S_i^{(k)}\}$ the iid service times of class-k outgoing calls with rate γ_k, and iid service times $\{S_i^{(k)}\}$ with rate $\mu_k = 1/\mathsf{E}S^{(k)}$, $k = 1,\dots,M$. If we denote by $A_k(t)$ the renewal process (number of calls) generated by the outgoing calls interval $[0,t)$, then the actual number of outgoing calls appeared in interval $[0,t)$, is (stochastically) equivalent to $A_k(I(t))$, where, recall, $I(t)$ is the empty time of server in $[0,t]$. In this case, keeping remaining notation, the balance equation (7) transforms to

$$V(t) = \sum_{i=1}^{A_0(t)} S_i + \sum_{k=1}^{M} \sum_{i=1}^{A_k(I(t))} S_i^{(k)} = S(t) + W_0(t) + t - I(t), \tag{18}$$

where now the remaining service time $S(t)$ may relate to any class-i outgoing call, $i = 1, \ldots, M$. Denote $\sigma_k = \gamma_k/\mu_k$, $k = 1, \ldots, M$. Then, as above in (7)–(11), we obtain that the stationary idle probability of the server is

$$P_0 = \frac{1 - \rho}{1 + \sum_{k=1}^{M} \sigma_k} = 1 - P_b. \tag{19}$$

For $M = 2$ this expression coincides with that is given in [9]. Also the same analysis as above gives the following expression for the stationary probability that server is busy by an incoming call $P_b^{(0)} = \rho$, which, as easy to check, is the same expression as in [9], p. 473. Finally, the summary work generated by class-k outgoing calls, in interval $[0, t)$ is

$$V_k(t) := \sum_{i=1}^{A_k(I(t))} S_i^{(k)}, \; k = 1, \ldots, M.$$

On the other hand, because the outgoing calls cannot be accumulated, the work $V_k(t)$ coincides with the busy time of the server devoted to class-k outgoing calls, in $[0, t)$. Then we obtain the following expression for the stationary probability $P_b^{(k)}$ that server is occupied by a class-k outgoing call (cf. (10)):

$$P_b^{(k)} = \lim_{t \to \infty} \frac{1}{t} \sum_{i=1}^{A_k(I(t))} S_i^{(k)} = \sigma_k P_0, \; k = 1, \ldots, M. \tag{20}$$

For $M = 2$, this expression is the same that is given in [9], p. 473. Then it follows from (20) that the stationary busy probability of server is

$$P_b = \sum_{k=0}^{M} P_b^{(k)} = \frac{\sum_{k=1}^{M} \sigma_k + \rho}{1 + \sum_{k=1}^{M} \sigma_k}, \tag{21}$$

and it is consistent with (19).

We emphasize that analysis above holds true for general renewal process of the incoming calls and general service time both incoming and outgoing calls. At the same time, exponentiality of the retrial times and the time between outgoing calls are being critically important for our analysis. It is reflected in the upper summation index $A_k(I(t))$ in (18) and in the term $D_0(L(t))$ in (13), where we, to obtain summary idle time $I(t)$ in general couple together *different periods* of the idle time of server. The same is done to obtain the summary time $L(t)$. It is well-known that such transformation preserves the property of Poisson process, that is $A_k(I(t))$ is equivalent to the number of the class-k outgoing calls appearing in interval $[0, I(t))$, provided the server devotes all its time to these calls.

3.3   Multiserver System

If the retrial system has m parallel stochastically identical servers, then previous analysis allows easily to obtain stationary idle probability $P_0^{(o)}$ of an arbitrary server. To this end, denote $I_i(t)$ the idle time of server i in interval $[0, t]$, $i = 1, \ldots, m$. Then the summary idle time in interval $[0, t]$ equals $I(t) = \sum_{i=1}^{m} I_i(t)$, and balance equation for (18) for the summary work transforms to the following balance equation

$$V(t) = \sum_{i=1}^{A_0(t)} S_i^{(1)} + \sum_{k=1}^{M} \sum_{i=1}^{A_k(I(t))} S_i^{(k)} = S(t) + W_0(t) + mt - I(t), \qquad (22)$$

where $S(t)$ is now the sum of the remaining service times in all servers, at instant t. Assume that the system is stable (positive recurrent), and denote the limit

$$\lim_{t \to \infty} \frac{I_i(t)}{t} =: \mathsf{P}_0^{(o)}, \ i = 1, \ldots, m,$$

which, because servers are identical, is independent of the server number i. Then we obtain in limit from (22) that

$$\rho + \sum_{k=1}^{m} \sigma_k \mathsf{P}_0^{(o)} = m - m\mathsf{P}_0^{(o)}, \qquad (23)$$

and thus the stationary idle probability of an arbitrary server equals

$$\mathsf{P}_0^{(o)} = \frac{m - \rho}{m + \sum_{k=1}^{M} \sigma_k}. \qquad (24)$$

Define $Q_i(t) = 1$ if server i is idle at instant t, and $Q_i(t) = 0$, otherwise, then

$$I_i(t) = \int_0^t Q_i(u)du, \ i = 1, \ldots, m,$$

and it follows that

$$\lim_{t \to \infty} \frac{I(t)}{t} = \lim_{t \to \infty} \frac{\sum_{i=1}^{m} I_i(t)}{t} = m\mathsf{P}_0^{(o)}$$

is the mean stationary number of empty servers.

4   Sufficient Stability Condition

In this section, we discuss in brief sufficient stability condition of the model. It has been obtained in [1] that condition

$$\rho + \frac{\lambda}{\mu_0}\rho + \frac{\lambda\gamma_1}{\mu_1\mu_0} < 1, \qquad (25)$$

is necessary for stability of two-way communication retrial system with one class of outgoing calls. Moreover, it has been noted that, visually, this condition is sufficient for stability as well. We now give an interpretation of this condition which allows to conjecture that (25) is a negative drift condition for the basic queue size process. This in turns implies that (25) is indeed stability criterion.

Consider first the model without outgoing calls. Evidently, because in this case $\sigma = 0$ then the necessary stability condition becomes (cf. (17))

$$\rho + \frac{\lambda}{\mu_0}\rho < 1. \tag{26}$$

To formulate negative drift condition for the workload, we must take into account that each new incoming call brings in the system the mean work equals $\mathsf{E}S$ and moreover, after each departure the server stays idle an exponential time, $\exp(\mu_0)$, with rate μ_0. In other words, within each interarrival time, $1/\mu_0$ is the lost capacity between departure and new attempt, *provided server is busy, and hence new call joins orbit*. It shows that the negative drift condition will be of the following form:

$$\lambda\Big(\mathsf{E}S + \mathsf{E}[I(Q > 0)\exp(\mu_0)]\Big) < 1, \tag{27}$$

where I is indicator function, Q is the stationary state of the server, in particular, $\mathsf{E}I(Q > 0) = \mathsf{P}_b = \rho$. It is seen that (27) coincides with (26).

Now we allow (one) class of outgoing calls (see Sect. 3.1). In this case, to obtain negative drift (sufficient stability condition), we must take into account the lost capacity of the server (or lost working time) because of the outgoing calls. To this end, we note that with the probability

$$q = \frac{\gamma_1}{\gamma_1 + \mu_0},$$

server is captured by an outgoing call for the (mean) service time $1/\mu_1$. Then, with probability q^2, it happens two times, etc. Thus, the total mean time devoted to outgoing calls between departure of a retrial call and the next successive attempt, equals $\gamma_1/(\mu_1\mu_0)$. Adding it to l.h.s. of (26), we obtain the following condition

$$\rho + \frac{\lambda}{\mu_0}\rho + \frac{\lambda\gamma_1}{\mu_1\mu_0} < 1, \tag{28}$$

which coincides with condition (17). For $M = 2$ classes we obtain, by analogy, the following negative drift condition:

$$\rho + \frac{\lambda}{\mu_0}\rho + \lambda\left[\frac{\gamma_1}{\mu_1(\gamma_2 + \mu_0)} + \frac{\gamma_2}{\mu_2(\gamma_1 + \mu_0)}\right] < 1. \tag{29}$$

Continuing in such a way, we can write down stability condition for M classes of the outgoing calls. However, we leave complete and strict analysis of stability criterion for this general case for a future work.

5 Conclusion

In this paper, we present a new regenerative proof of some recent steady-state performance results for a two-way communication retrial queueing model with constant retrial rate [1,8,9] obtained earlier by Kolmogorov equations approach. Also we deduce necessary stability condition and discuss sufficient condition as well. Moreover, some generalizations of the model are also considered.

Acknowledgement. The research of EM is partly supported by Russian Foundation for Basic Research, projects 18-07-00156 and 18-07-00147 and also by the Institute of Applied Mathematical Research, Karelian Research Centre RAS. The research of TP is partially supported by University of Tsukuba Basic Research Support Program Type A.

References

1. Aissani, A., Phung-Duc, T.: Profiting the idleness in single server system with orbit-queue. In: Proceedings of VALUETOOLS 2017, 5–7 December 2017, Venice, Italy (2017)
2. Aksin, Z., Armony, M., Mehrotra, V.: The modern call center: a multidisciplinary perspective on operations management research. Prod. Oper. Manag. **16**, 665–688 (2007)
3. Asmussen, S.: Applied Probability and Queues. Stochastic Modelling and Applied Probability, 2nd edn. Springer, New York (2003). https://doi.org/10.1007/b97236
4. Feller, W.: An Introduction to Probability Theory, vol. 2. Wiley, New York (1971)
5. Morozov, E.: The tightness in the ergodic analysis of regenerative queueing processes. Queueing Syst. **27**, 179203 (1997)
6. Morozov, E., Delgado, R.: Stability analysis of regenerative queues. Autom. Remote Control **70**, 1977–1991 (2009)
7. Morozov, E., Phung-Duc, T.: Stability analysis of a multiclass retrial system with classical retrial policy. Perform. Eval. **112**, 15–26 (2017)
8. Phung-Duc, T., Rogiest, W., Takahashi, Y., Bruneel, H.: Retrial queues with balanced call blending: analysis of single-server and multiserver model. Annal. Oper. Res. **239**, 429–449 (2016)
9. Sakurai, H., Phung-Duc, T.: Two-way communication retrial queues with multiple types of outgoing calls. TOP **23**, 466–492 (2015)
10. Smith, W.L.: Regenerative stochastic processes. Proc. R. Soc. Ser. A **232**, 6–31 (1955)

A Retrial Queueing System with a Variable Number of Active Servers: Dynamic Manpower Planning in a Call Center

Rein Nobel[(✉)]

Department of Econometrics and Operations Research, Vrije Universiteit
Amsterdam, Amsterdam, The Netherlands
r.d.nobel@vu.nl

Abstract. A retrial queueing model is considered with Poisson input
and an unlimited number of servers. At any epoch only a finite number
of the servers are active, the others are called dormant. An active server
is always in one of two possible states, idle or busy. When upon arrival
of a customer at least one of the active servers is idle, the newly arrived
customer goes into service immediately, making the idle server busy.
When at an arrival epoch all active servers are busy, the decision must
be made to send the newly arrived customer into orbit, or to activate a
dormant server for immediate service of the arrived customer. Customers
in orbit try to reenter the system after an exponentially distributed retrial
time. At service completion epochs the decision must be made to keep
the newly become idle server active, or to make this server dormant.
The service times of the customers are independent and have a Coxian-2
distribution. Given specific costs for activating servers, keeping servers
active and a holding cost for customers staying in orbit, the problem is
when to activate and shut down servers in order to minimize the long-
run average cost per unit time. Using Markov decision theory an efficient
algorithm is discussed for calculating an optimal policy.

Keywords: Retrial model · Semi-Markov decision model
Fictitious decision epochs

1 Introduction

In recent years we have seen a considerable increase in the number of call centers. Both private companies and governmental institutions use these centers for
answering questions from their customers. As a consequence, a lot of research
has been undertaken to study the random behavior of these call centers. Not surprisingly, queueing theory plays a dominant role in this research. Starting with
Erlang's Loss model, many papers have been written in which besides lost calls,
also retrials and abandonments have been incorporated. For a tutorial overview

Y. Takahashi et al. (Eds.): QTNA 2018, LNCS 10932, pp. 33–47, 2018.
https://doi.org/10.1007/978-3-319-93736-6_3

we refer to [4,6] and the references therein. A nice introductory paper on abandonments is [8] in which the so-called Palm/Erlang-A model is discussed. For
the impact of retrials on call center performance we refer to [1]. The main topic
in call center research is to find a balance between service quality, expressed
e.g. in waiting-time characteristics, and the cost of operation, expressed e.g. in
the number of active servers. Hence, the so-called 'staffing problem' is a central topic in most of the call center literature. Formulated in the terminology of
queueing theory, this problem can be described as follows. Given all the relevant
parameters for some multi-server queueing model in which customers have the
option to abandon the system after not having been served within some random time, and/or to retry entrance to the system some random time later after
an unsuccessful arrival, the question is how many active servers (agents) must
be available to guarantee a required balance between service quality and operational cost. The given parameters of the queueing model include the arrival
rate, the service rate, the abandonment rate and the retrial rate. The design
parameter is the number of agents. Taking a model with *all* parameter values
fixed as a starting point, most papers give a descriptive analysis for the steady-
state behavior of the system. Due to their complexity these models often do not
allow for a practically feasible exact solution. This is *a fortiori* the case for the
transient behavior of the systems and/or when parameters are time-dependent.
To cope with this intractability, all kinds of approximations are considered, such
as fluid and diffusion approximations, e.g. see [7]. We will not give an extensive
overview of all the research on call centers done so far, but the point we want to
make is that most of this research is descriptive in character: the steady-state or
transient behavior of the models is studied for a set of *given* parameters. Much
less attention has been dedicated to finding dynamic operational policies as a
solution for the staffing problem.

So, instead of giving a descriptive analysis of some queueing model with a
fixed number of servers, we propose to study the staffing problem as a dynamic
optimization problem: let the number of active servers depend on the current
congestion of the system, expressed in the number of busy servers and the number of waiting customers, and increase or decrease the number of active servers
depending on instantaneous changes of the congestion as a consequence of an
arrival, a departure or an abandonment. To pursue this idea of dynamic man-
power planning, we propose to study a retrial multi-server queueing model with
an adaptable number of servers. To limit the calculational burden, we do not
consider abandonments in this paper, but we want to underline that abandonments can be easily incorporated in the model, if one wishes to do so. For this
retrial model, to be described in detail below, we will use Markov decision theory
to calculate an operating policy, for which a subtle balance between the costs of
congestion and the operational costs is minimized.

In a standard retrial model (see [2,3] for monographs on retrial queues) customers who find all servers busy try to enter the system some time later. We say
that the customer goes into orbit. Nowadays it is very common that the system
knows at any time how many customers are in orbit (we can simply register

unsuccessful calls). So, this information can be used in the determination of the number of active servers. In our model the number of servers is unlimited but at any moment only a finite number is active (the others are called dormant), and this number is under control of the management of the system. Hence, we consider a multi-server retrial queueing system with a controllable number of active servers, who can be idle or busy. When upon arrival of a customer no idle server is available, a choice must be made to activate a dormant server, or send the newly arrived customer into orbit. Upon a service completion it must be decided to shut down an active idle server (i.e. make him dormant), or keep him active for possible new arrivals. Of course, these decisions must be guided by some optimization criterion, i.e. a cost structure must be introduced in the model. Given this cost structure, the problem is to find the strategy for activating and shutting down servers, for which the long-run average cost per unit time is minimal. This strategy is a so-called stationary dynamic strategy, i.e. the decisions prescribed by the strategy take into account all the relevant information available at the decision epochs, not more and not less, in other words, the decisions are based on the complete state description of the system. By choosing a specific stochastic structure with respect to the probability distributions involved, we can describe our problem in terms of a semi-Markov decision model. A straightforward application of a standard algorithm from Markov decision theory is not feasible here, due to the large state space. By introducing so-called *fictitious* decision epochs we will show how to overcome this obstacle.

In Sect. 2 the queueing model is described in detail. Section 3 describes a semi-Markov decision model and the value-iteration algorithm to calculate the optimal policy for which the long-run average cost per unit time is minimal. In Sect. 4 some numerical results are given.

2   Description of the Model

We consider a queueing model with retrials and a controllable number of active servers. The number of servers is unlimited, but at any epoch only a finite number of servers is active, either idle or busy. The non-active servers are called dormant. For idle servers linear operating costs α per server per unit time are incurred, whereas for busy servers these costs are γ per server per unit time ($\gamma > \alpha$). Customers arrive at the system according to a Poisson process with rate λ. Each customer requires a service time denoted by the generic variable S, and the service times of different customers are independent. To observe the influence of the variability of the service times S (expressed via the squared coefficient of variation C_S^2) on the optimal strategies and the long-run average cost, we model the service times S with a Coxian-2 distribution with parameters b, μ_1 and μ_2 with $0 < b < 1$ and $\mu_1 < \mu_2$. We recall here that this says that S is, with probability b, distributed as a sum of two independent exponential phases, say S_1 and S_2 with mean $1/\mu_1$ and $1/\mu_2$, respectively, and with probability $1 - b$, S is distributed as one exponential phase S_1 with mean $1/\mu_1$. As we will see in the next section, the Coxian-2 distribution is a very convenient choice for the service times, due to the memoryless property of the exponential phases. Also it is easy to fit a Coxian-2 distribution when only the first two moments are given (see [11] for further details). When upon arrival of a customer at least one idle server is available, the customer immediately starts its service, reducing the number of idle servers by one. When no idle server is present, either the newly arrived customer goes into orbit, or a dormant server is activated to serve the customer immediately. Activating a dormant server requires a set-up cost K. Customers in orbit try to enter the system again after an exponentially distributed retrial time R with mean $1/\nu$. The different retrial times are independent. For customers in orbit linear holding costs h per customer per unit time are incurred. At service completion epochs the choice must be made between keeping the server active (in the idle state) or making him dormant. So, the question is when to activate a dormant server upon an arrival and when to shut down an active server upon a departure, in order to minimize the long-run average cost per unit time.

To calculate this optimal policy, in the next section we will formulate the model as a *semi-Markov decision model*. To avoid the problem of an infinite state space we will give our analysis for a *truncated* model, i.e. we limit the number of available servers to a finite number C and the maximum number of customers allowed to be in orbit will be taken M. A customer in orbit is considered to stay in orbit until he is accepted for service. So, to complete the description of the truncated model, we must specify precisely what to do upon arrivals and departures in the boundary situations:

- New arrivals who find M customers in orbit and less than C servers busy will always be accepted,
- An arrival from orbit will always be accepted when M customers are in orbit and less than C servers are busy,
- New arrivals who find M customers in orbit and C servers busy are rejected,
- An arrival from orbit will stay in orbit when upon arrival C servers are busy,
- New arrivals who find C servers busy and less than M customers in orbit are always sent to orbit,

By taking C and M sufficiently large the fraction of customers which will be rejected is negligible, so our numerical results will be valid for the non-truncated model as well.

3   The Semi-Markov Decision Model

We assume that the reader is acquainted with the concepts of Markov decision theory (see [10,11] for thorough introductions to this subject), so we will not give an extensive description of the building blocks of a semi-Markov decision model. We just recall that we have to specify a state space $\mathcal{S}$, action sets $\mathcal{A}(s)$ for each state $s \in \mathcal{S}$, a matrix of transition probabilities $p[s'|s, a]$ for $s', s \in \mathcal{S}$ and $a \in \mathcal{A}(s)$, expected one-step costs $\eta[s, a]$ for $s \in \mathcal{S}$ and $a \in \mathcal{A}(s)$, and the expected sojourn times $\tau[s, a]$ for each state $s \in \mathcal{S}$ and $a \in \mathcal{A}(s)$. All these building blocks will now be specified for the controllable queueing model described in the previous section. To describe a semi-Markov decision model with a sparse matrix of transition probabilities, it is very convenient to introduce so-called fictitious decision epochs (see [9,11]). According to the model description the only decision epochs are the arrival epochs at which no idle server is present and the epochs of service completion. To guarantee a sparse matrix of transition probabilities, we include all arrival epochs, and phase completion epochs of service times as decision epochs as well. At these latter epochs no real decision is made, we just leave the system as it is. We denote this 'no action' decision by 0. At service completion epochs two decisions are possible, shut down the server who has just become idle (denoted by −1), or leave the system as it is (again denoted by 0). At arrival epochs we leave the system as it is, when an idle server is available. A real decision has to be made only when all active servers are busy. Then we can send the newly arrived job into orbit (denoted by 0), or we can activate a dormant server (denoted by 1). At this point it will be clear why our choice for Coxian service times is so convenient: due to the exponential phases of the service time it is sufficient to know whether a service is in its first phase or in its second phase, and this enables a simple description of the state of the system, as we will see next. In fact we can introduce the following state description at the decision epochs,

$$(i, j_1, j_2, k, e), \quad i = 0, 1, \ldots; \quad j_1, j_2 = 0, 1, \ldots; k = 0, 1, \ldots; \quad e = 0, 1, -1, -2,$$

with the following interpretation

- i is the number of idle servers,
- j_1 is the number busy servers in the first service phase,
- j_2 is the number busy servers in the second service phase,
- k is the number of jobs in orbit,
- e describes the type of event that occurred: $e = 0$ denotes a new arrival (from the Poisson stream), $e = 1$ denotes an arrival from the orbit, $e = -1$ stands for a phase completion of an ongoing service time, which leads to the next phase of this service time, and $e = -2$ stands for the completion of a service time.

We emphasize that the numbers i, j_1, j_2, k always refer to the numbers just *after* the 'event' e has occurred, but *before* the decision is made. Specifically, a customer in orbit is considered to stay in orbit until he is accepted for service. Notice that not all combinations (i, j_1, j_2, k, e) refer to real states, e.g. the states $(0, j_1, j_2, k, -2)$ do not exist because upon a departure $(e = -2)$ at least one server must be idle.

Next, we will specify the elements of the semi-Markov decision model

$$(\mathcal{S}, \{\mathcal{A}(s), s \in \mathcal{S}\}, \{\tau[s, a], s \in \mathcal{S}, a \in \mathcal{A}(s)\}, \{\eta[s, a], s \in \mathcal{S}, a \in \mathcal{A}(s)\},$$

$$\{p[t|s, a], s, t \in \mathcal{S}, a \in \mathcal{A}(s)\}),$$

which describes the retrial queueing model with a variable number of active servers. As already indicated above, the state space $\mathcal{S}$ is taken as

$$\mathcal{S} = \{(i, j_1, j_2, k, e) \mid i = 0, 1, \ldots; j_1, j_2 = 0, 1, \ldots; i+j_1+j_2 \leq C; k = 0, 1, \ldots, M;$$

$$e = 0, 1, -1, -2\}.$$

The action sets $\mathcal{A}(s)$ are very simple. At each service completion epoch the decision must be made to shut down an idle server or to leave the system as it is. When at an arrival epoch no idle servers are available the decision must be made to switch on a dormant server or send the newly arrived job into orbit. Because we do not allow more than M customers in orbit, we always accept an arriving customer when M customers are in orbit, unless the number of active servers is C. In the latter case we reject primary customers from the Poisson stream and leave the arriving customers from orbit in orbit until the number of busy servers has dropped below C. These remarks lead to the following action sets.

$$
\begin{aligned}
\mathcal{A}(i, j_1, j_2, k, -2) &= \{0, -1\}, & & i = 1, 2, \ldots; \; j_1, j_2, k = 0, 1, 2, \ldots, \\
\mathcal{A}(0, j_1, j_2, k, e) &= \{0, 1\}, & & e = 0, 1; \; j_1, j_2 = 0, 1, \ldots; \; j_1 + j_2 < C; \\
& & & k = e, 1, 2, \ldots, M-1, \\
\mathcal{A}(0, j_1, j_2, M, e) &= \{1\}, & & e = 0, 1; \; j_1, j_2 = 0, 1, \ldots; \; j_1 + j_2 < C, \\
\mathcal{A}(0, j_1, j_2, k, e) &= \{0\}, & & e = 0, 1; \; j_1, j_2 = 0, 1, \ldots; \; j_1 + j_2 = C; \\
& & & k = e, 1, 2, \ldots, M, \\
\mathcal{A}(i, j_1, j_2, k, e) &= \{0\}, & & e = 0, 1; \; i = 1, 2, \ldots; \; j_1, j_2 = 0, 1, \ldots; \\
& & & k = e, 1, 2, \ldots, \\
\mathcal{A}(i, j_1, j_2, k, -1) &= \{0\}, & & i, j_1, j_2, k = 0, 1, \ldots.
\end{aligned}
$$

For the one-step transition probabilities $p[s'|s, a]$, denoting the conditional probability that, given action a is taken in state s, at the next decision epoch the state is s', we first consider the real decision epochs, i.e. the service completion epochs, and the arrival epochs when no idle server is available. Firstly, we give the one-step transition probabilities given that a service completion has occurred. So the decision a is either 0 (keep all idle servers active) or -1 (switch off an idle server).

$$p[(i + a, j_1, j_2, k, 0) \mid (i, j_1, j_2, k, -2), a] = \frac{\lambda}{\lambda + j_1\mu_1 + j_2\mu_2 + k\nu},$$

$$p[(i + a, j_1, j_2, k, 1) \mid (i, j_1, j_2, k, -2), a] = \frac{k\nu}{\lambda + j_1\mu_1 + j_2\mu_2 + k\nu},$$

$$p[(i + a, j_1 - 1, j_2 + 1, k, -1) \mid (i, j_1, j_2, k, -2), a] = \frac{bj_1\mu_1}{\lambda + j_1\mu_1 + j_2\mu_2 + k\nu},$$

$$p[(i + a + 1, j_1 - 1, j_2, k, -2) \mid (i, j_1, j_2, k, -2), a] = \frac{(1 - b)j_1\mu_1}{\lambda + j_1\mu_1 + j_2\mu_2 + k\nu},$$

$$p[(i + a + 1, j_1, j_2 - 1, k, -2) \mid (i, j_1, j_2, k, -2), a] = \frac{j_2\mu_2}{\lambda + j_1\mu_1 + j_2\mu_2 + k\nu}.$$

Of course, the third and fourth transition is only possible when $j_1 > 0$, and the last transition requires that $j_2 > 0$.

Next, we write down these probabilities given that an arrival has taken place, a primary arrival $(e = 0)$ or an arrival from orbit $(e = 1)$, and no idle servers are present. Now, the decision $a = 0$ stands for 'send (keep) he arrived customer (in) to orbit' and $a = 1$ denotes 'switch on a dormant server'.

$$p[(0, j_1 + a, j_2, k + 1 - a - e, 0) \mid (0, j_1, j_2, k, e), a] = \frac{\lambda}{\mathcal{N}(j_1, j_2, k, e, a)},$$

$$p[(0, j_1 + a, j_2, k + 1 - a - e, 1) \mid (0, j_1, j_2, k, e), a] = \frac{(k + 1 - a - e)\nu}{\mathcal{N}(j_1, j_2, k, e, a)},$$

$$p[(0, j_1 + a - 1, j_2 + 1, k + 1 - a - e, -1) \mid (0, j_1, j_2, k, e), a] = \frac{b(j_1 + a)\mu_1}{\mathcal{N}(j_1, j_2, k, e, a)},$$

$$p[(1, j_1 + a - 1, j_2, k + 1 - a - e, -2) \mid (0, j_1, j_2, k, e), a] = \frac{(1 - b)(j_1 + a)\mu_1}{\mathcal{N}(j_1, j_2, k, e, a)},$$

$$p[(1, j_1 + a, j_2 - 1, k + 1 - a - e, -2) \mid (0, j_1, j_2, k, e), a] = \frac{j_2\mu_2}{\mathcal{N}(j_1, j_2, k, e, a)},$$

where $\mathcal{N}(j_1, j_2, k, e, a) := \lambda + (j_1 + a)\mu_1 + j_2\mu_2 + (k + 1 - a - e)\nu$ is the common denominator.

Similarly, we can treat the fictitious decision epochs, the arrival epochs with $i > 0$ idle servers available, and the phase completion epochs. The only decision is now 0 (leave the system as it is).

$$p[(i - 1, j_1 + 1, j_2, k - e, 0) \mid (i, j_1, j_2, k, e), 0] = \frac{\lambda}{\lambda + (j_1 + 1)\mu_1 + j_2\mu_2 + (k - e)\nu},$$

$$p[(i - 1, j_1 + 1, j_2, k - e, 1) \mid (i, j_1, j_2, k, e), 0] = \frac{(k - e)\nu}{\lambda + (j_1 + 1)\mu_1 + j_2\mu_2 + (k - e)\nu},$$

$$p[(i - 1, j_1, j_2 + 1, k - e, -1) \mid (i, j_1, j_2, k, e), 0] = \frac{b(j_1 + 1)\mu_1}{\lambda + (j_1 + 1)\mu_1 + j_2\mu_2 + (k - e)\nu},$$

$$p[(i, j_1, j_2, k - e, -2) \mid (i, j_1, j_2, k, e), 0] = \frac{(1 - b)(j_1 + 1)\mu_1}{\lambda + (j_1 + 1)\mu_1 + j_2\mu_2 + (k - e)\nu},$$

$$p[(i, j_1 + 1, j_2 - 1, k - e, -2) \mid (i, j_1, j_2, k, e), 0] = \frac{j_2\mu_2}{\lambda + (j_1 + 1)\mu_1 + j_2\mu_2 + (k - e)\nu},$$

$$p[(i, j_1, j_2, k, 0) \mid (i, j_1, j_2, k, -1), 0] = \frac{\lambda}{\lambda + j_1\mu_1 + j_2\mu_2 + k\nu},$$

$$p[(i, j_1, j_2, k, 1) \mid (i, j_1, j_2, k, -1), 0] = \frac{k\nu}{\lambda + j_1\mu_1 + j_2\mu_2 + k\nu},$$

$$p[(i, j_1 - 1, j_2 + 1, k, -1) \mid (i, j_1, j_2, k, -1), 0] = \frac{bj_1\mu_1}{\lambda + j_1\mu_1 + j_2\mu_2 + k\nu},$$

$$p[(i + 1, j_1 - 1, j_2, k, -2) \mid (i, j_1, j_2, k, -1), 0] = \frac{(1 - b)j_1\mu_1}{\lambda + j_1\mu_1 + j_2\mu_2 + k\nu},$$

$$p[(i + 1, j_1, j_2 - 1, k, -2) \mid (i, j_1, j_2, k, -1), 0] = \frac{j_2\mu_2}{\lambda + j_1\mu_1 + j_2\mu_2 + k\nu}.$$

Next, we consider the boundary cases, i.e. the number of customers in orbit is M and/or the number of busy servers is C. First, we look at arrivals finding M

customers in orbit and less than C servers busy ($j_1 + j_2 < C$). As stated above in this case we always accept a new customer.

$$p[(0, j_1 + 1, j_2, M - e, 0) \mid (0, j_1, j_2, M, e), 1] = \frac{\lambda}{\lambda + (j_1 + 1)\mu_1 + j_2\mu_2 + (M - e)\nu},$$

$$p[(0, j_1 + 1, j_2, M - e, 1) \mid (0, j_1, j_2, M, e), 1] = \frac{(M - e)\nu}{\lambda + (j_1 + 1)\mu_1 + j_2\mu_2 + (M - e)\nu},$$

$$p[(0, j_1, j_2 + 1, M - e, -1) \mid (0, j_1, j_2, M, e), 1] = \frac{b(j_1 + 1)\mu_1}{\lambda + (j_1 + 1)\mu_1 + j_2\mu_2 + (M - e)\nu},$$

$$p[(1, j_1, j_2, M - e, -2) \mid (0, j_1, j_2, M, e), 1] = \frac{(1 - b)(j_1 + 1)\mu_1}{\lambda + (j_1 + 1)\mu_1 + j_2\mu_2 + (M - e)\nu},$$

$$p[(1, j_1, j_2 - 1, M - e, -2) \mid (0, j_1, j_2, M, e), 1] = \frac{j_2\mu_2}{\lambda + (j_1 + 1)\mu_1 + j_2\mu_2 + (M - e)\nu},$$

Now we consider arrivals who find C servers busy (so $j_1 + j_2 = C$) and less than M customers in orbit. They are always sent into orbit.

$$p[(0, j_1, j_2, k + 1 - e, 0) \mid (0, j_1, j_2, k, e), 0] = \frac{\lambda}{\lambda + j_1\mu_1 + j_2\mu_2 + (k + 1 - e)\nu},$$

$$p[(0, j_1 + 1, j_2, k + 1 - e, 1) \mid (0, j_1, j_2, k, e), 0] = \frac{(k + 1 - e)\nu}{\lambda + j_1\mu_1 + j_2\mu_2 + (k + 1 - e)\nu},$$

$$p[(0, j_1 - 1, j_2 + 1, k + 1 - e, -1) \mid (0, j_1, j_2, k, e), 0] = \frac{bj_1\mu_1}{\lambda + j_1\mu_1 + j_2\mu_2 + (k + 1 - e)\nu},$$

$$p[(1, j_1 - 1, j_2, k + 1 - e, -2) \mid (0, j_1, j_2, k, e), 0] = \frac{(1 - b)j_1\mu_1}{\lambda + j_1\mu_1 + j_2\mu_2 + (k + 1 - e)\nu},$$

$$p[(1, j_1, j_2 - 1, k + 1 - e, -2) \mid (0, j_1, j_2, k, e), 0] = \frac{j_2\mu_2}{\lambda + j_1\mu_1 + j_2\mu_2 + (k + 1 - e)\nu}.$$

Finally, we look at arrival epochs when C servers are busy and M customers are in orbit. Then new arrival are rejected and arrivals from orbit stay in orbit. So we get, ($j_1 + j_2 = C$, $e = 0, 1$)

$$p[(0, j_1, j_2, M, 0) \mid (0, j_1, j_2, M, e), 0] = \frac{\lambda}{\lambda + j_1\mu_1 + j_2\mu_2 + M\nu},$$

$$p[(0, j_1, j_2, M, 1) \mid (0, j_1, j_2, M, e), 0] = \frac{M\nu}{\lambda + j_1\mu_1 + j_2\mu_2 + M\nu},$$

$$p[(0, j_1 - 1, j_2 + 1, M, -1) \mid (0, j_1, j_2, M, e), 0] = \frac{bj_1\mu_1}{\lambda + j_1\mu_1 + j_2\mu_2 + M\nu},$$

$$p[(1, j_1 - 1, j_2, M, -2) \mid (0, j_1, j_2, M, e), 0] = \frac{(1 - b)j_1\mu_1}{\lambda + j_1\mu_1 + j_2\mu_2 + M\nu},$$

$$p[(1, j_1, j_2 - 1, M, -2) \mid (0, j_1, j_2, M, e), 0] = \frac{j_2\mu_2}{\lambda + j_1\mu_1 + j_2\mu_2 + M\nu}.$$

Let us next consider the $\tau[s, a]$, i.e. the expected time until the next decision epoch given that in state s action a is chosen.

$$\tau[(i, j_1, j_2, k, e), 0] = \frac{1}{\lambda + (j_1 + 1)\mu_1 + j_2\mu_2 + (k - e)\nu}, \quad i = 1, 2, \ldots,$$

$$\tau[(0, j_1, j_2, k, e), a] = \frac{1}{\lambda + (j_1 + a)\mu_1 + j_2\mu_2 + (k + 1 - a - e)\nu}, \quad a = 0, 1,$$

$$\tau[(i, j_1, j_2, k, -1), 0] = \frac{1}{\lambda + j_1\mu_1 + j_2\mu_2 + k\nu}, \quad i = 0, 1, 2, \ldots,$$

$$\tau[(i, j_1, j_2, k, -2), a] = \frac{1}{\lambda + j_1\mu_1 + j_2\mu_2 + k\nu}, \quad i = 1, 2, \ldots, \quad a = 0, -1.$$

To complete the formulation of the Markov-decision model we must specify the costs $\eta[s, a]$, i.e., the total expected costs incurred until the next decision epoch when in state s action a is taken. We give a few examples.

$$\eta[(i, j_1, j_2, k, e), 0] = \frac{(i - 1)\alpha + (j_1 + 1 + j_2)\gamma + (k - e)h}{\lambda + (j_1 + 1)\mu_1 + j_2\mu_2 + (k - e)\nu}, \quad i = 1, 2, \ldots,$$

$$\eta[(0, j_1, j_2, k, e), a] = aK + \frac{(j_1 + a + j_2)\gamma + (k + 1 - a - e)h}{\lambda + (j_1 + a)\mu_1 + j_2\mu_2 + (k + 1 - a - e)\nu},$$

$$\eta[(i, j_1, j_2, k, -1), 0] = \frac{i\alpha + (j_1 + j_2)\gamma + kh}{\lambda + j_1\mu_1 + j_2\mu_2 + k\nu}, \quad i = 0, 1, 2, \ldots,$$

$$\eta[(i, j_1, j_2, k, -2), a] = \frac{(i + a)\alpha + (j_1 + j_2)\gamma + kh}{\lambda + j_1\mu_1 + j_2\mu_2 + k\nu}, \quad i = 1, 2, \ldots; \quad a = 0, -1.$$

Once all the elements of the Markov-decision model are known, we can use the value-iteration algorithm to calculate the optimal switching strategy. We give the formulation of the algorithm in general terms (see [11] for any further details). First choose a positive number τ with $\tau \leq \min_{s,a} \tau_s(a)$ and a tolerance number ϵ, e.g., $\epsilon = 10^{-6}$.

INIT. For all $s \in \mathcal{S}$, choose nonnegative numbers $W_0(s)$ with $W_0(s) \leq \min_a\{\eta[s, a]/\tau[s, a]\}$. Let $n := 1$.
LOOP. For all $s \in \mathcal{S}$, calculate

$$W_n(s) = \min_{a \in \mathcal{A}(s)} \left[\frac{\eta[s, a]}{\tau[s, a]} + \frac{\tau}{\tau[s, a]} \sum_{t \in \mathcal{S}} p[t \mid s, a] W_{n-1}(t) + \left\{ 1 - \frac{\tau}{\tau[s, a]} \right\} W_{n-1}(s) \right],$$

and let $D_n(s) \in \mathcal{A}(s)$ be the action that minimizes the right-hand side.
EVAL. Compute the bounds,

$$m_n = \min_{s \in \mathcal{S}}\{W_n(s) - W_{n-1}(s)\}, \quad M_n = \max_{s \in \mathcal{S}}\{W_n(s) - W_{n-1}(s)\}.$$

TEST. If $M_n - m_n \leq \epsilon m_n$ then STOP with the resulting policy D_n, else $n := n + 1$ and go to **LOOP.**

This algorithm returns after say n iterations a stationary policy D_n^* that minimizes the long-run average costs per unit time. The (approximate) minimal average costs is calculated as $g^* = (m_n + M_n)/2$.

4   Numerical Results

In this section we will present some numerical results. Because there are many parameters which can be varied we must make a selection. To start, we have chosen to keep the arrival rate and the mean service time constant and we vary only the retrial rate, and the squared coefficient of variation of the service time (we used Gamma normalization for fitting the parameters of the Coxian-2 distribution; see [11] for the details how to choose the parameters b, μ_1 and μ_2 to guarantee a given mean and squared coefficient of variation). Because the mathematical state-description is more detailed than any reasonable physical state-description, we present a natural heuristic policy, with the corresponding cost, besides the optimal solution. To explain the heuristics, notice that the mathematical state-description in our model contains the phases of the ongoing services which cannot be observed physically. Because in practice we only observe the number of idle servers i, the number of busy servers j, and, by registration, the number in orbit k (in other words on occurrence of the event e the *physical* state is (i, j, k, e)), we must base our decisions on this information for all different *mathematical* states (i, j_1, j_2, k, e) with $j_1 + j_2 = j$. So, we are forced to select one decision in all these latter states, whereas the (mathematically) optimal policy may prescribe different decisions for these states. We have chosen a kind of *democratic* heuristic rule, defined as follows. When in the majority of the states (i, j_1, j_2, k, e) decision a is the optimal decision, then we prescribe this decision in all corresponding physical states (i, j, k, e) with $j = j_1 + j_2$). In Tables 1 and 2 we present the minimal cost and the corresponding heuristic cost for the following parameter values,

$$\lambda = 3, \quad E[S] = 2, \quad h = 10, \quad \alpha = 20, \quad \gamma = 25, \quad K = 500.$$

In Table 1 the holding cost for staying in the orbit $h = 10$ and in Table 2 we have chosen $h = 1$. We vary the retrial rate ν and the squared coefficient of variation

Table 1. Minimal and heuristic cost for $\lambda = 3$, $E[S] = 2$, $h = 10$, $\alpha = 20$, $\gamma = 25$, $K = 500$.

$\nu \backslash C_S^2$	0.6	0.8	1	2	4	8
0.25	228.53	229.03	229.53	231.76	234.61	234.55
	228.53	229.03	229.53	231.81	235.15	239.44
0.5	216.13	217.18	218.14	221.87	225.43	226.36
	216.16	217.19	218.14	222.00	226.73	231.20
1	207.94	209.05	210.08	214.34	218.90	220.38
	207.96	209.05	210.08	214.44	220.06	226.17
2	203.63	204.71	205.72	210.02	214.94	216.32
	203.66	204.71	205.72	210.12	215.84	222.98
4	200.72	202.35	203.38	207.62	212.65	213.72
	200.74	202.36	203.38	207.69	214.24	220.36

Table 2. Minimal and heuristic cost for $\lambda = 3$, $E[S] = 2$, $h = 1$, $\alpha = 20$, $\gamma = 25$, $K = 500$.

$\nu \backslash C_S^2$	0.6	0.8	1	2	4	8
0.25	192.15	193.24	194.22	197.97	201.89	203.98
	192.19	193.26	194.22	198.20	202.72	205.43
0.5	184.09	185.37	186.55	191.14	196.16	199.01
	184.13	185.40	186.55	191.40	196.80	200.73
1	179.61	180.87	182.05	186.94	192.56	195.57
	179.64	180.87	182.05	187.11	193.08	197.30
2	177.36	178.53	179.67	184.56	190.43	193.36
	177.38	178.54	179.67	184.69	190.90	195.31
4	176.05	177.33	178.42	183.25	189.15	192.02
	176.12	177.33	178.42	183.37	189.82	193.98

Table 3. Primary arivals strategy $C_S^2 = 1$, $\nu = 4$, $h = 1$, $\alpha = 20$, $\gamma = 25$, $K = 500$.

act \ orb	0	1	2	3	4	5	6	7	8	9	10	11	12	13	14	15	16	17	18	19	20
0	1	1	1	1	1	1	1	1	1	1	1	1	1	1	1	1	1	1	1	1	1
1	0	1	1	1	1	1	1	1	1	1	1	1	1	1	1	1	1	1	1	1	1
2	0	1	1	1	1	1	1	1	1	1	1	1	1	1	1	1	1	1	1	1	1
3	0	0	1	1	1	1	1	1	1	1	1	1	1	1	1	1	1	1	1	1	1
4	0	0	0	1	1	1	1	1	1	1	1	1	1	1	1	1	1	1	1	1	1
5	0	0	0	0	0	0	1	1	1	1	1	1	1	1	1	1	1	1	1	1	1
6	0	0	0	0	0	0	0	0	0	0	0	0	0	0	1	1	1	1	1	1	1
7	0	0	0	0	0	0	0	0	0	0	0	0	0	0	0	0	0	0	1	1	1
8	0	0	0	0	0	0	0	0	0	0	0	0	0	0	0	0	0	0	0	0	1
9	0	0	0	0	0	0	0	0	0	0	0	0	0	0	0	0	0	0	0	0	1
10	0	0	0	0	0	0	0	0	0	0	0	0	0	0	0	0	0	0	0	0	1
11	0	0	0	0	0	0	0	0	0	0	0	0	0	0	0	0	0	0	0	0	1
12	0	0	0	0	0	0	0	0	0	0	0	0	0	0	0	0	0	0	0	0	1
13	0	0	0	0	0	0	0	0	0	0	0	0	0	0	0	0	0	0	0	0	1
14	0	0	0	0	0	0	0	0	0	0	0	0	0	0	0	0	0	0	0	0	1
15	0	0	0	0	0	0	0	0	0	0	0	0	0	0	0	0	0	0	0	0	1
16	0	0	0	0	0	0	0	0	0	0	0	0	0	0	0	0	0	0	0	0	1
17	0	0	0	0	0	0	0	0	0	0	0	0	0	0	0	0	0	0	0	0	1
18	0	0	0	0	0	0	0	0	0	0	0	0	0	0	0	0	0	0	0	0	1
19	0	0	0	0	0	0	0	0	0	0	0	0	0	0	0	0	0	0	0	0	1
20	0	0	0	0	0	0	0	0	0	0	0	0	0	0	0	0	0	0	0	0	0

Table 4. Departures strategy $C_S^2 = 1$, $\nu = 4$, $h = 1$, $\alpha = 20$, $\gamma = 25$, $K = 500$.

act \ orb	0	1	2	3	4	5	6	7	8	9	10	11	12	13	14	15	16	17	18	19	20
0	0	0	0	0	0	0	0	0	0	0	0	0	0	0	0	0	0	0	0	0	0
1	0	0	0	0	0	0	0	0	0	0	0	0	0	0	0	0	0	0	0	0	0
2	0	0	0	0	0	0	0	0	0	0	0	0	0	0	0	0	0	0	0	0	0
3	0	0	0	0	0	0	0	0	0	0	0	0	0	0	0	0	0	0	0	0	0
4	0	0	0	0	0	0	0	0	0	0	0	0	0	0	0	0	0	0	0	0	0
5	0	0	0	0	0	0	0	0	0	0	0	0	0	0	0	0	0	0	0	0	0
6	0	0	0	0	0	0	0	0	0	0	0	0	0	0	0	0	0	0	0	0	0
7	0	0	0	0	0	0	0	0	0	0	0	0	0	0	0	0	0	0	0	0	0
8	1	1	1	1	1	1	1	0	0	0	0	0	0	0	0	0	0	0	0	0	0
9	1	1	1	1	1	1	1	1	1	1	0	0	0	0	0	0	0	0	0	0	0
10	1	1	1	1	1	1	1	1	1	1	1	1	1	0	0	0	0	0	0	0	0
11	1	1	1	1	1	1	1	1	1	1	1	1	1	1	1	0	0	0	0	0	0
12	1	1	1	1	1	1	1	1	1	1	1	1	1	1	1	1	1	0	0	0	0
13	1	1	1	1	1	1	1	1	1	1	1	1	1	1	1	1	1	1	0	0	0
14	1	1	1	1	1	1	1	1	1	1	1	1	1	1	1	1	1	1	1	0	0
15	1	1	1	1	1	1	1	1	1	1	1	1	1	1	1	1	1	1	1	1	0
16	1	1	1	1	1	1	1	1	1	1	1	1	1	1	1	1	1	1	1	1	1
17	1	1	1	1	1	1	1	1	1	1	1	1	1	1	1	1	1	1	1	1	1
18	1	1	1	1	1	1	1	1	1	1	1	1	1	1	1	1	1	1	1	1	1
19	1	1	1	1	1	1	1	1	1	1	1	1	1	1	1	1	1	1	1	1	1
20	1	1	1	1	1	1	1	1	1	1	1	1	1	1	1	1	1	1	1	1	1

of the service time C_S^2. Notice that the difference between the optimal cost and the (democratic) heuristic cost is negligible for $C_s^2 \leq 1$. This difference turns out to be significant only for high holding costs and very irregular service times.

To give an idea of the 'form' of the strategies for turning on and off servers we present both the primary arrival strategy and one departure strategy for a specific choice of the parameters. In Tables 3 and 4 we present these optimal (heuristic) strategies for exponential service and in Tables 5 and 6 for very irregular service times ($C_S^2 = 8$). In these tables the number of customers in orbit is presented horizontally and the number of active servers vertically. Notice that in the tables for the arrival strategies the active servers are all busy (otherwise there is nothing to decide), but for the departure strategies the decisions are not based on the number of active servers alone; we also need to know how many servers are busy. So, for each specific number of active servers, say i, to be complete we should present i rows, i.e. one row for each possible number of idle servers. To avoid such an overwhelming amount of information in one table, in Tables 5 and 6 we have made the choice to show only the decisions for the

Table 5. Primary Arrivals STRATEGY $C_S^2 = 8$, $\nu = 4$, $h = 1$, $\alpha = 20$, $\gamma = 25$, $K = 500$.

act \ orb	0	1	2	3	4	5	6	7	8	9	10	11	12	13	14	15	16	17	18	19	20
0	0	1	1	1	1	1	1	1	1	1	1	1	1	1	1	1	1	1	1	1	1
1	0	0	0	1	1	1	1	1	1	1	1	1	1	1	1	1	1	1	1	1	1
2	0	0	0	0	1	1	1	1	1	1	1	1	1	1	1	1	1	1	1	1	1
3	0	0	0	0	0	0	0	0	0	0	1	1	1	1	1	1	1	1	1	1	1
4	0	0	0	0	0	0	0	0	0	0	0	0	1	1	1	1	1	1	1	1	1
5	0	0	0	0	0	0	0	0	0	0	0	0	0	0	0	0	0	0	1	1	1
6	0	0	0	0	0	0	0	0	0	0	0	0	0	0	0	0	0	0	0	1	1
7	0	0	0	0	0	0	0	0	0	0	0	0	0	0	0	0	0	0	0	0	1
8	0	0	0	0	0	0	0	0	0	0	0	0	0	0	0	0	0	0	0	0	1
9	0	0	0	0	0	0	0	0	0	0	0	0	0	0	0	0	0	0	0	0	1
10	0	0	0	0	0	0	0	0	0	0	0	0	0	0	0	0	0	0	0	0	1
11	0	0	0	0	0	0	0	0	0	0	0	0	0	0	0	0	0	0	0	0	1
12	0	0	0	0	0	0	0	0	0	0	0	0	0	0	0	0	0	0	0	0	1
13	0	0	0	0	0	0	0	0	0	0	0	0	0	0	0	0	0	0	0	0	1
14	0	0	0	0	0	0	0	0	0	0	0	0	0	0	0	0	0	0	0	0	1
15	0	0	0	0	0	0	0	0	0	0	0	0	0	0	0	0	0	0	0	0	1
16	0	0	0	0	0	0	0	0	0	0	0	0	0	0	0	0	0	0	0	0	1
17	0	0	0	0	0	0	0	0	0	0	0	0	0	0	0	0	0	0	0	0	1
18	0	0	0	0	0	0	0	0	0	0	0	0	0	0	0	0	0	0	0	0	1
19	0	0	0	0	0	0	0	0	0	0	0	0	0	0	0	0	0	0	0	0	1
20	0	0	0	0	0	0	0	0	0	0	0	0	0	0	0	0	0	0	0	0	0

situation that half of the active servers is idle. Finally, in these tables a '0' stands for 'leave the system as it is', so for the arrival strategies: send the newly arrived customer into orbit, and for the departure strategies 'keep the server just become idle active', and a '1' for 'turning on a dormant server' (arrival) and 'turning off an idle server' (departure). We see from these tables that the policies are rather insensitive for the squared coefficient of variation of the service time, whereas the associated costs are quite different (see Table 2, for $C_S^2 = 1$, i.e. exponentially distributed service times, the minimal costs are 178.42, and for $C_S^2 = 8$, *ceteris paribus*, 193.98). This fact, that optimal policies are rather robust for the variability of the service time, is a well-known phenomenon in the literature on controlled queueing systems. But we can add as an interesting conclusion that the optimal strategies become more conservative as C_S^2 becomes larger, i.e. the

Table 6. DEPARTURES STRATEGY $C_S^2 = 8$, $\nu = 4$, $h = 1$, $\alpha = 20$, $\gamma = 25$, $K = 500$.

act \ orb	0	1	2	3	4	5	6	7	8	9	10	11	12	13	14	15	16	17	18	19	20
0	0	0	0	0	0	0	0	0	0	0	0	0	0	0	0	0	0	0	0	0	0
1	0	0	0	0	0	0	0	0	0	0	0	0	0	0	0	0	0	0	0	0	0
2	0	0	0	0	0	0	0	0	0	0	0	0	0	0	0	0	0	0	0	0	0
3	0	0	0	0	0	0	0	0	0	0	0	0	0	0	0	0	0	0	0	0	0
4	0	0	0	0	0	0	0	0	0	0	0	0	0	0	0	0	0	0	0	0	0
5	0	0	0	0	0	0	0	0	0	0	0	0	0	0	0	0	0	0	0	0	0
6	0	0	0	0	0	0	0	0	0	0	0	0	0	0	0	0	0	0	0	0	0
7	0	0	0	0	0	0	0	0	0	0	0	0	0	0	0	0	0	0	0	0	0
8	1	1	0	0	0	0	0	0	0	0	0	0	0	0	0	0	0	0	0	0	0
9	1	1	1	1	0	0	0	0	0	0	0	0	0	0	0	0	0	0	0	0	0
10	1	1	1	1	1	1	1	1	1	0	0	0	0	0	0	0	0	0	0	0	0
11	1	1	1	1	1	1	1	1	1	1	0	0	0	0	0	0	0	0	0	0	0
12	1	1	1	1	1	1	1	1	1	1	1	1	1	1	1	0	0	0	0	0	0
13	1	1	1	1	1	1	1	1	1	1	1	1	1	1	1	1	0	0	0	0	0
14	1	1	1	1	1	1	1	1	1	1	1	1	1	1	1	1	1	1	1	1	0
15	1	1	1	1	1	1	1	1	1	1	1	1	1	1	1	1	1	1	1	1	0
16	1	1	1	1	1	1	1	1	1	1	1	1	1	1	1	1	1	1	1	1	1
17	1	1	1	1	1	1	1	1	1	1	1	1	1	1	1	1	1	1	1	1	1
18	1	1	1	1	1	1	1	1	1	1	1	1	1	1	1	1	1	1	1	1	1
19	1	1	1	1	1	1	1	1	1	1	1	1	1	1	1	1	1	1	1	1	1
20	1	1	1	1	1	1	1	1	1	1	1	1	1	1	1	1	1	1	1	1	1

system is less eager to switch on/off a server for larger values of C_S^2 than for smaller values of C_S^2.

References

1. Aguir, S., Karaesmen, F., Aksin, Z., Chauvet, F.: The impact of retrials on call center performance. OR Spectr. **26**, 353–376 (2004)
2. Artalejo, J.R., Gómez-Corral, A.: Retrial Queueing Systems. Springer, Heidelberg (2008). https://doi.org/10.1007/978-3-540-78725-9
3. Falin, G., Templeton, J.: Retrial Queues. Chapman & Hall, London (1997)
4. Gans, N., Koole, G., Mandelbaum, A.: Telephone call centers: tutorial, review and research prospects. Manuf. Serv. Oper. Manag. **5**, 79–141 (2003)
5. Garnett, O., Mandelbaum, A., Reiman, M.: Designing a call center with impatient customers. Manuf. Serv. Oper. Manag. **4**, 208–227 (2002)

6. Koole, G., Mandelbaum, A.: Queueing models of call centers, an introduction. Ann. Oper. Res. **113**, 41–59 (2002)

7. Mandelbaum, A., Massey, W., Reiman, M., Stolyar, A., Rider, B.: Queue lengths and waiting times for multiserver queues with abandonment and retrials. Telecommun. Syst. **21**, 149–171 (2002)

8. Mandelbaum, A., Zeltyn, S.: Service engineering in action: the Palm/Erlang-A queue, with applications to call centers. In: Spath, D., Fähnrich, K.P. (eds.) Advances in Services Innovations, pp. 17–45. Springer, Heidelberg (2007). https:// doi.org/10.1007/978-3-540-29860-1_2

9. Nobel, R.D., Bekker, R.: Optimal control for two queues with one switching server. In: Proceedings of the National Conference on Mathematical and Computational Models, PSG College of Technology, Coimbatore, India, pp. 21–33, 27–28 December 2001

10. Puterman, M.L.: Markov Decision Processes. Wiley, New York (1994)

11. Tijms, H.C.: A First Course in Stochastic Models. Wiley, New York (2003)

Queueing Models with Variants of Service

Controllable Capacity Queue with Synchronous Constant Service Time and Loss

Zsolt Saffer[1]([✉]), Karl Grill[1], and Wuyi Yue[2]

[1] Institute of Statistics and Mathematical Methods in Economics,
Vienna University of Technology, Vienna, Austria
{zsolt.saffer,karl.grill}@tuwien.ac.at
[2] Department of Intelligence and Informatics, Konan University,
Kobe 658-8501, Japan
yue@konan-u.ac.jp

Abstract. In this paper we analyze a queue, in which the number of active servers can be controlled by means of probabilities specifying the dependency of the number of active servers on the actual number of customers and the number of active servers. We call this queue as controllable capacity queue. The service time is constant and the concurrently served customers are served in syncronized manner. The active number of servers can be incremented, decremented or kept unchanged at the ends of service time according to the given probabilities. The system has no buffer for long-term customer waiting, it is a loss system. Such system could be relevant in modeling Machine-To-Machine communication systems, in which the resources are limited.

We provide explicit form results for the joint and marginal distributions of the number of servers and the number of customers on PGF level. We give the condition of the stability and also provide the expressions of the most important system measures including the mean stationary number of servers, the mean stationary number of customers and the blocking probability.

Keywords: Queueing theory · Control of queues
State dependent number of servers · Variable number of servers

1 Introduction

Controlling of available server capacity is a natural demand in the areas, which can be modelled by queueing systems, like modern telecommunication systems, manufacturing systems, call centers. More recent application areas include base station aggregating traffic in Machine-To-Machine communication [1] or Intelligent transportation systems [2].

Capacity can be modelled on various manner, like time slice, variable number of work units the system can perform, service rate of model with exponential

© Springer International Publishing AG, part of Springer Nature 2018
Y. Takahashi et al. (Eds.): QTNA 2018, LNCS 10932, pp. 51–63, 2018.
https://doi.org/10.1007/978-3-319-93736-6_4

service time or by the number of active servers. The idea of using capacity as time slice is motivated by Data Frame based Media Access Control (MAC) protocols, in which a scheduling mechanism allocates only a portion of time to the uplink data transmission of a target subscriber station. The dynamical change of capacity based on this approach has been analyzed in [3]. The approach for capacity as variable number of work units the system can perform has been presented by Bruneel et al. in [4], in which they also provide an analysis of dynamical change of such capacity. The above works consider the dynamical change of capacity, but not the control of it.

Several works can be found in the literature on control of service rate as capacity, which put the capacity modeling in the context of M/M queueing system. A fundamental work on such service rate control is [5], in which the capacity control is realized by letting the service rate depend on the number of customers in the system. This dependency specifies the service rate to be proportional to the power of the number of customers. In the paper [6] the authors study M/M/s queueing system, in which the number of servers can vary between a lower limit and an upper limit. In the work [7] the authors study a control schema, in which the number of active servers increases when the queue grows by k customers and decreases accordingly. To the best knowledge of the authors no work is available on controlling the number of servers in a queueing system with non-exponentially distributed service time.

In this paper in contrast to the above references we consider the control of the number of servers in a queue with constant service time. The number of active servers depends on the actual number of customers and the number of active servers and it can be controlled by means of probabilities specifying these dependency. We call this queue as controllable capacity queue. The concurrently served customers are served in syncronized manner. The active number of servers can be incremented, decremented or kept unchanged at the ends of service time with some tunable probabilities specifying the dependency of the number of active servers on the actual number of customers in the system and the actual number of active servers. This controllable capacity queue has no buffer for long-term customer waiting, therefore it is a loss system. Such system could be relevant in modeling network nodes with microcontroller having limited resources, like e.g. in Machine-To-Machine communication systems or in Wireless Sensor Networks.

We characterize the model by a bivariate discrete-time Markov chain (DTMC) embedded at end epochs of the customer service. Utilizing both the tridiagonal structure in the one and the M/D like structure in the other dimensions of the DTMC, we perform a stationary analysis of the model by applying probability generating function (PGF) techniques. We derive explicit form results for the joint and marginal distributions of the number of servers and the number of customers on PGF level. We give the condition of the stability and also provide the expressions of the most important system measures including the mean stationary number of servers, the mean stationary number of customers and the blocking probability.

The rest of the paper is organized as follows. In Sect. 2 we give the detailed description of the model and establish the embedded DTMC of the system. Preliminary stationary results are provided in Sect. 3. The major part of the stationary analysis follows in Sect. 4, which includes the derivation of the PGFs of the joint and marginal distributions and the stability criterion. Section 5 presents the expression of the system measures. Final remarks closes the work in Sect. 6.

2   Model and Notation

In this section we give a detailed description of the model and then we establish an embedded DTMC for this model.

2.1   Model Description

We consider a queue with controllable capacity, where the capacity is realized by the number of servers. Customers arrive to the system according to Poisson process with parameter $0 < \lambda < \infty$. The customer service time is constant and it is denoted by s, for which $0 < s < \infty$ holds. The concurrently served customers are served in syncronized manner, which means that their service starts at the same time and s time later ends also at the same time. The number of servers can be changed at each time when the service period finishes. It can be incremented, decremented or kept unchanged with some probabilities depending on the actual number of customers and servers. The number of servers, m is unlimited, but there is always at least one active server, also in the case when there is no customer present in the system, i.e. $1 \leq m$. The model has no buffer for long-term customer waiting, i.e. waiting longer than one customer service time, therefore it is a loss system. However the system includes temporary buffers for making the accumulation of the customers during the actual service period possible. If the system has m active servers then the maximum number of customers allowed to be accumulated is $m + 1$ due to the potential increment of the number of servers at the end of the actual service period. We assume that the arrival process, the customer service time and the change of the number of servers are mutually independent.

Let $q(m, k)$ be the probability of decrementing the number of servers at end of service period, when there are m active servers and the number of customers in the system is k, for $m \geq 2$ and $k \geq 0$. Similarly let $r(m, k)$ stand for the probability of incrementing the number of servers at end of service period, when there are m active servers and the number of customers in the system is k, for $m \geq 1$ and $k \geq 0$. These probabilities are given in the form

$$q(m, k) = (1 - \delta^m)\gamma^k, \qquad (1)$$
$$r(m, k) = \alpha(1 - q(m, k)) = \alpha \left(1 - (1 - \delta^m)\gamma^k\right), \quad k = 0, \ldots, m,$$
$$\text{where } 0 < \alpha \leq 1, \quad 0 \leq \gamma \leq 1, \quad 0 \leq \delta < 1.$$

This rule provides a simple control on the probabilities of decrementing, incrementing and keeping unchanged the number of servers in terms of parameters

γ, δ and α. The construction of the forms in (1) ensures that $0 \leq q(m, k) \leq 1$ and $0 \leq r(m, k) \leq 1$, i.e. they are probabilities. The parameter γ implements a control objective depending on the number of customer. The higher the number of customers and closer to m, the lower (higher) the probability of decrementing (incrementing) the number of servers. On the other hand the parameter δ implements a control objective depending on the number of servers. The higher the number of servers, the more important to decrease or keeping unchanged the number of servers and hence higher (lower) the probability of decrementing (incrementing) the number of servers. The parameter α controls the probability of no change in the number of servers, which is given as

$$1 - (r(m, k) + q(m, k)) = 1 - (\alpha + (1 - \alpha)q(m, k)) = (1 - \alpha)(1 - q(m, k)),$$

which is always between 0 and 1, since $0 \leq \alpha, q(m, k) \leq 1$.

We assume that the model is stable. For the stability condition see Subsect. 4.2. In the following we call the model in short form as controllable capacity queue.

2.2   Formulating an Embedded DTMC

Let $t^d(n)$ stands for the end of the n-th customer service, for $n \geq 1$. Let $t^{d+}(n)$ and $t^{d-}(n)$ be the epoch just after and just before the end of the n-th customer service, respectively. Furthermore let $t^{d--}(n)$ be the epoch just before the epoch $t^{d-}(n)$. The exact order of the events in the system related to the end of the n-th customer service can be seen in Fig. 1.

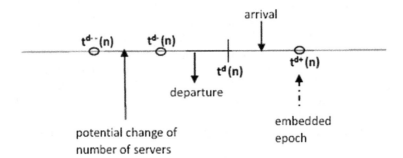

Fig. 1. The order of events related to the end of the n-th customer service

Remark 1. This order would correspond to early arrival system (EAS) in the discrete-time queueing model, in which this order of arrival and departure would eliminate the need for having temporary buffer.

We consider the evolution of the system at the $t^{d+}(n)$ epochs, $n \geq 1$. Let $M(t)$ and $N(t)$ stand for the number of servers and the number of customers in the system at time $t > 0$. Then the process $(M(t^{d+}(n)), N(t^{d+}(n)), n \geq 1)$

is a bivariate Markov chain embedded at epochs just after the end of the n-th customer service. The transition probability matrix of this embedded DTMC, $\mathbf{P}$ can be given by the help of block matrices. Let the main index of matrix $\mathbf{P}$ be the number of servers (m) and let the index inside of the block matrices be the number of customers (n). Then the transition probability matrix $\mathbf{P}$ can be given in terms of the model parameters as

$$
\mathbf{P} = \begin{pmatrix}
(\mathbf{I} - \mathbf{D}_1^I)\mathbf{A}_{1,1} & \mathbf{D}_1^I\mathbf{A}_{1,2} & \mathbf{0} & \mathbf{0} & \cdots \\
\mathbf{D}_2^D\mathbf{A}_{2,1} & (\mathbf{I} - \mathbf{D}_2^D - \mathbf{D}_2^I)\mathbf{A}_{2,2} & \mathbf{D}_2^I\mathbf{A}_{2,3} & \mathbf{0} & \cdots \\
\mathbf{0} & \mathbf{D}_3^D\mathbf{A}_{3,2} & (\mathbf{I} - \mathbf{D}_3^D - \mathbf{D}_3^I)\mathbf{A}_{3,3} & \mathbf{D}_3^I\mathbf{A}_{3,4} & \cdots \\
\vdots & \vdots & \vdots & \vdots & \ddots
\end{pmatrix}. \quad (2)
$$

The matrices D_m^D for $m \geq 2$ and D_m^I for $m \geq 1$ are $(m+1) \times (m+1)$ diagonal block matrices, which are given as

$$
\mathbf{D}_m^D = \begin{pmatrix}
q(m,0) & 0 & 0 & \cdots \\
0 & q(m,1) & 0 & \cdots \\
\vdots & \vdots & \vdots & \ddots \\
0 & 0 & \cdots & q(m,m)
\end{pmatrix}, \quad
\mathbf{D}_m^I = \begin{pmatrix}
r(m,0) & 0 & 0 & \cdots \\
0 & r(m,1) & 0 & \cdots \\
\vdots & \vdots & \vdots & \ddots \\
0 & 0 & \cdots & r(m,m)
\end{pmatrix}.
$$

The matrices $\mathbf{A}_{m,m+i}$ for $m \geq 1$ and $i = -1, 0, 1$ are $(m+1) \times (m+1+i)$ block matrices representing the probabilities of the number of arrivals during the transition $m \Rightarrow (m+i)$ in the number of servers, i.e. during which the number of servers changes from m to $(m+i)$ and they are given as

$$
\mathbf{A}_{m,m} = \begin{pmatrix}
a_0 & \cdots & a_{m-1} & a_m^+ \\
a_0 & \cdots & a_{m-1} & a_m^+ \\
\vdots & \ddots & \vdots & \vdots \\
a_0 & \cdots & a_{m-1} & a_m^+
\end{pmatrix},
$$

$$
\mathbf{A}_{m,m-1} = \begin{pmatrix}
a_0 & \cdots & a_{m-2} & a_{m-1}^+ \\
a_0 & \cdots & a_{m-2} & a_{m-1}^+ \\
\vdots & \ddots & \vdots & \vdots \\
a_0 & \cdots & a_{m-2} & a_{m-1}^+
\end{pmatrix}, \quad
\mathbf{A}_{m,m+1} = \begin{pmatrix}
a_0 & \cdots & a_m & a_{m+1}^+ \\
a_0 & \cdots & a_m & a_{m+1}^+ \\
\vdots & \ddots & \vdots & \vdots \\
a_0 & \cdots & a_m & a_{m+1}^+
\end{pmatrix},
$$

where the probabilities a_k for $k \geq 0$ and a_m^+ for $m \geq 1$ are given as

$$
a_k = P\left(k \text{ arrivals during service time of length } s\right) = \frac{(\lambda s)^k}{k!}e^{-\lambda s},
$$

$$
a_m^+ = \sum_{k=m}^{\infty} a_k.
$$

We define the PGF $a_m(z)$ as

$$
a_m(z) = \sum_{k=0}^{m-1} a_k z^k + a_m^+ z^m, \quad |z| \leq 1, m \geq 1.
$$

For the PGF $a(z)$, $a^{(k)}$ denotes its k-th derivative at $z = 1$ for $k \geq 1$, i.e., $a^{(k)} = \frac{d^k}{dz^k} a(z)|_{z=1}$.

3   Preliminary Stationary Results

In this section we provide several preliminary results, which

- on one hand give some insight into the behavior of the system and
- on other hand constitute the base for the derivation of the main results.

Let $p_{m,k}(t^{d+}(n))$ be the probability that the system is in state (m, k) at time $t^{d+}(n)$, i.e. there are m active servers and k customers in the system, for $m \geq 1$, $k \geq 0$ and $n \geq 1$. We define the stationary probabilities $p_{m,k}$ as

$$p_{m,k} = \lim_{n \to \infty} p_{m,k}(t^{d+}(n)), \quad m \geq 1, k \geq 0.$$

We also define the stationary probability vectors $\mathbf{p}_m$ as

$$\mathbf{p}_m = (p_{m,0}, \ldots, p_{m,m}), \quad m \geq 1.$$

Based on the transition probability matrix (2) the stationary equations of the embedded DTMC can be formulated as

$$\mathbf{p}_1(\mathbf{I} - \mathbf{D}_1^{\mathbf{I}})\mathbf{A}_{11} + \mathbf{p}_2\mathbf{D}_2^{\mathbf{D}}\mathbf{A}_{21} = \mathbf{p}_1, \tag{3}$$

$$\mathbf{p}_{m-1}\mathbf{D}_{m-1}^{\mathbf{I}}\mathbf{A}_{m-1,m} + \mathbf{p}_m(\mathbf{I} - \mathbf{D}_m^{\mathbf{D}} - \mathbf{D}_m^{\mathbf{I}})\mathbf{A}_{m,m}$$
$$+ \mathbf{p}_{m+1}\mathbf{D}_{m+1}^{\mathbf{D}}\mathbf{A}_{m+1,m} = \mathbf{p}_m, \quad m \geq 2.$$

Let the vector $\mathbf{z}_m$ be defined as

$$\mathbf{z}_m = (1, z, \ldots, z^m), \quad |z| \leq 1, m \geq 1.$$

Multiplying the equations of (3) by $\mathbf{z}_m$ from right gives the scalar form stationary equations of the embedded DTMC as

$$\sum_{k=0}^{1} p_{1,k}(1 - r(1,k))a_1(z) + \sum_{k=0}^{2} p_{2,k}q(2,k)a_1(z) = \sum_{k=0}^{1} p_{1,k}z^k. \tag{4}$$

$$\sum_{k=0}^{m-1} p_{m-1,k}r(m-1,k)a_m(z) + \sum_{k=0}^{m} p_{m,k}(1 - q(m,k) - r(m,k))a_m(z)$$
$$+ \sum_{k=0}^{m+1} p_{m+1,k}q(m+1,k)a_m(z) = \sum_{k=0}^{m} p_{m,k}z^k, \quad m \geq 2.$$

Proposition 1 (Balance equation). *A balance equation holds in the stable controllable capacity queue as*

$$\sum_{k=0}^{m+1} p_{m+1,k}q(m+1,k) = \sum_{k=0}^{m} p_{m,k}r(m,k), \quad m \geq 1. \tag{5}$$

Proof. Setting $z = 1$ in the second relation of (4) and rearrangement yields

$$\sum_{k=0}^{m+1} p_{m+1,k} q(m+1,k) - \sum_{k=0}^{m} p_{m,k} q(m,k)$$

$$= \sum_{k=0}^{m} p_{m,k} r(m,k) - \sum_{k=0}^{m-1} p_{m-1,k} r(m-1,k), \quad m \geq 2. \tag{6}$$

Summing up (6) for $m = 2, \ldots, i$ gives

$$\sum_{m=2}^{i} \left(\sum_{k=0}^{m+1} p_{m+1,k} q(m+1,k) - \sum_{k=0}^{m} p_{m,k} q(m,k) \right)$$

$$= \sum_{m=2}^{i} \left(\sum_{k=0}^{m} p_{m,k} r(m,k) - \sum_{k=0}^{m-1} p_{m-1,k} r(m-1,k) \right), \quad m \geq 2,$$

which can be rearranged as

$$\sum_{k=0}^{i+1} p_{i+1,k} q(i+1,k) - \sum_{k=0}^{2} p_{2,k} q(2,k) = \sum_{m=2}^{i} p_{i,k} r(i,k) - \sum_{k=0}^{1} p_{1,k} r(1,k). \tag{7}$$

Expressing $\sum_{k=0}^{2} p_{2,k} q(2,k)$ from the first relation of (4) for $z = 1$, applying it in (7), rearranging it and observing that it holds also for $m = 1$ results in the statement of the proposition. □

Remark 2. The reason of existence such a kind of balance equation for the controllable capacity queue model is that the structure of the scalar stationary equations, (4) are similar to that of the birth-death process.

We introduce the following PGFs

$$p_m(z) = \sum_{k=0}^{m} p_{m,k} z^k, \quad |z| \leq 1, m \geq 1.$$

$$p(u, z) = \sum_{m=1}^{\infty} p_m(z) u^{m-1} = \sum_{m=1}^{\infty} \sum_{k=0}^{m} p_{m,k} z^k u^{m-1}, \quad |u| \leq 1, |z| \leq 1.$$

Theorem 1. *In the stable controllable capacity queue the joint distribution of the number of servers and the number of customers is determined by the marginal distribution of the servers as*

$$p_m(z) = p_m(1) a_m(z), \quad m \geq 1. \tag{8}$$

Proof. Setting $z = 0$ in (4) and rearrangement gives

$$\sum_{k=0}^{1} p_{1,k} \left(1 - r(1,k)\right) + \sum_{k=0}^{2} p_{2,k} q(2,k) = \frac{p_{1,0}}{a_0}, \tag{9}$$

$$\sum_{k=0}^{m-1} p_{m-1,k} r(m-1,k) + \sum_{k=0}^{m} p_{m,k} \left(1 - q(m,k) - r(m,k)\right)$$

$$+ \sum_{k=0}^{m+1} p_{m+1,k} q(m+1,k) = \frac{p_{m,0}}{a_0}, \quad m \geq 2.$$

Comparing (9) to (4) yields

$$\frac{p_{m,0}}{a_0} a_m(z) = \sum_{k=0}^{m} p_{m,k} z^k, \quad m \geq 1. \tag{10}$$

Setting $z = 1$ in (10) gives

$$\frac{p_{m,0}}{a_0} = \sum_{k=0}^{m} p_{m,k}, \quad m \geq 1. \tag{11}$$

Now combining (11) and (10) results in

$$\sum_{k=0}^{m} p_{m,k} z^k = \sum_{k=0}^{m} p_{m,k} a_m(z), \quad m \geq 1. \tag{12}$$

The statement of the theorem comes by applying the PGF notations in (12). □

4   The Stationary Distribution of the Number of Servers and the Number of Customers

In this section we present the main result, the joint stationary distribution of the number of servers and the number of customers. Afterwards we derive the stability criterion of the model and the marginal stationary distributions.

4.1   The Joint Stationary Distribution of the Number of Servers and the Number of Customers

Lemma 1. *The following relations hold in the controllable capacity queue*

$$\sum_{k=0}^{m} p_{m,k} q(m,k) = (1 - \delta^m) p_m(\gamma), \tag{13}$$

$$\sum_{k=0}^{m} p_{m,k} r(m,k) = \alpha \left(p_m(1) - (1 - \delta^m) p_m(\gamma)\right), \quad m \geq 1.$$

Proof. Applying the defining relations (1) in $\sum_{k=0}^{m} p_{m,k} q(m,k)$ and $\sum_{k=0}^{m} p_{m,k} r(m,k)$ gives

$$\sum_{k=0}^{m} p_{m,k} q(m,k) = \sum_{k=0}^{m} p_{m,k}(1-\delta^m)\gamma^k = (1-\delta^m)\sum_{k=0}^{m} p_{m,k}\gamma^k, \quad (14)$$

$$\sum_{k=0}^{m} p_{m,k} r(m,k) = \sum_{k=0}^{m} p_{m,k}\alpha\left(1-(1-\delta^m)\gamma^k\right)$$

$$= \alpha \sum_{k=0}^{m} p_{m,k} - \alpha(1-\delta^m)\sum_{k=0}^{m} p_{m,k}\gamma^k.$$

The lemma comes by applying the PGF notations in (14). □

Theorem 2. *The PGF of the joint distribution of the number of servers and number of customers, with respect to the number of customers in the stable controllable capacity queue is given as*

$$p_m(z) = \frac{\frac{a_m(z)}{a_m(\gamma)} \prod_{i=1}^{m-1} \frac{\alpha\left(1-(1-\delta^i)a_i(\gamma)\right)}{(1-\delta^{i+1})a_i(\gamma)}}{\sum_{m=1}^{\infty} \frac{1}{a_m(\gamma)} \prod_{i=1}^{m-1} \frac{\alpha(1-(1-\delta^i)a_i(\gamma))}{(1-\delta^{i+1})a_i(\gamma)}}, \quad m \geq 1. \quad (15)$$

Proof. Applying the relations of Lemma 1 in (5) gives

$$(1-\delta^{m+1})p_{m+1}(\gamma) = \alpha\left(p_m(1) - (1-\delta^m)p_m(\gamma)\right), \quad m \geq 1. \quad (16)$$

Multiplying (16) by $a_m(z)$ and applying (8) in it leads to

$$(1-\delta^{m+1})p_{m+1}(\gamma)a_m(z) = \alpha\left(p_m(1)a_m(z) - (1-\delta^m)p_m(\gamma)a_m(z)\right) \quad (17)$$
$$= \alpha\left(p_m(z) - (1-\delta^m)p_m(\gamma)a_m(z)\right), \quad m \geq 1.$$

Utilizing that γ lies in the convergence range of z, $|z| \leq 1$, we set $z = \gamma$ in (17), which yields

$$(1-\delta^{m+1})p_{m+1}(\gamma)a_m(\gamma) = \alpha\left(p_m(\gamma) - (1-\delta^m)p_m(\gamma)a_m(\gamma)\right), \quad m \geq 1. \quad (18)$$

Rearranging (18) gives the expression of $p_{m+1}(\gamma)$ in terms of $p_m(\gamma)$ as

$$p_{m+1}(\gamma) = \frac{\alpha\left(1-(1-\delta^m)a_m(\gamma)\right)}{(1-\delta^{m+1})a_m(\gamma)}p_m(\gamma). \quad (19)$$

Solving (19) by recursive substitution for $m = 1, \ldots$ results in the expression of $p_m(\gamma)$ in terms of the unknown constant $p_1(\gamma)$ as

$$p_m(\gamma) = p_1(\gamma) \prod_{i=1}^{m-1} \frac{\alpha\left(1-(1-\delta^i)a_i(\gamma)\right)}{(1-\delta^{i+1})a_i(\gamma)}. \quad (20)$$

Expressing $p_m(1)$ from (16) gives

$$p_m(1) = (1 - \delta^m)p_m(\gamma) + \frac{(1 - \delta^{m+1})p_{m+1}(\gamma)}{\alpha}, \quad m \geq 1. \tag{21}$$

Now we apply (20) in (21) and perform rearrangement, which leads to

$$p_m(1) = p_1(\gamma) *$$
$$\left((1 - \delta^m) \prod_{i=1}^{m-1} \frac{\alpha \left(1 - (1 - \delta^i)a_i(\gamma)\right)}{(1 - \delta^{i+1})a_i(\gamma)} + \frac{(1 - \delta^{m+1})}{\alpha} \prod_{i=1}^{m} \frac{\alpha \left(1 - (1 - \delta^i)a_i(\gamma)\right)}{(1 - \delta^{i+1})a_i(\gamma)} \right)$$
$$= p_1(\gamma) \left((1 - \delta^m) + \frac{(1 - \delta^{m+1})}{\alpha} \frac{\alpha \left(1 - (1 - \delta^m)a_m(\gamma)\right)}{(1 - \delta^{m+1})a_m(\gamma)} \right) \prod_{i=1}^{m-1} \frac{\alpha \left(1 - (1 - \delta^i)a_i(\gamma)\right)}{(1 - \delta^{i+1})a_i(\gamma)}$$
$$= p_1(\gamma) \frac{1}{a_m(\gamma)} \prod_{i=1}^{m-1} \frac{\alpha \left(1 - (1 - \delta^i)a_i(\gamma)\right)}{(1 - \delta^{i+1})a_i(\gamma)}, \quad m \geq 1. \tag{22}$$

Applying (22) in the statement of Theorem 1 results in the expression of the PGF of the joint distribution of the number of servers and number of customers, with respect to the number of customers as

$$p_m(z) = p_1(\gamma) \frac{a_m(z)}{a_m(\gamma)} \prod_{i=1}^{m-1} \frac{\alpha \left(1 - (1 - \delta^i)a_i(\gamma)\right)}{(1 - \delta^{i+1})a_i(\gamma)}, \quad m \geq 1. \tag{23}$$

The unknown constant $p_1(\gamma)$ in (23) can be determined from the normalization condition as

$$1 = \sum_{m=1}^{\infty} p_m(1) = p_1(\gamma) \sum_{m=1}^{\infty} \frac{1}{a_m(\gamma)} \prod_{i=1}^{m-1} \frac{\alpha \left(1 - (1 - \delta^i)a_i(\gamma)\right)}{(1 - \delta^{i+1})a_i(\gamma)},$$

from which

$$p_1(\gamma) = \frac{1}{\sum_{m=1}^{\infty} \frac{1}{a_m(\gamma)} \prod_{i=1}^{m-1} \frac{\alpha(1-(1-\delta^i)a_i(\gamma))}{(1-\delta^{i+1})a_i(\gamma)}}. \tag{24}$$

The statement of the theorem comes by applying (24) in (23). □

Corollary 1. *The joint PGF of the joint distribution of the number of servers and number of customers in the stable controllable capacity queue is given as*

$$p(u, z) = \frac{\sum_{m=1}^{\infty} u^{m-1} \frac{a_m(z)}{a_m(\gamma)} \prod_{i=1}^{m-1} \frac{\alpha\left(1-(1-\delta^i)a_i(\gamma)\right)}{(1-\delta^{i+1})a_i(\gamma)}}{\sum_{m=1}^{\infty} \frac{1}{a_m(\gamma)} \prod_{i=1}^{m-1} \frac{\alpha(1-(1-\delta^i)a_i(\gamma))}{(1-\delta^{i+1})a_i(\gamma)}}. \tag{25}$$

Proof. The statement of the corollary comes by taking the PGF of (15) with respect to m. □

4.2   Stability

Corollary 2. *The necessary and sufficient condition of the stability of the controllable capacity queue is given as*

$$\sum_{m=1}^{\infty} \frac{1}{a_m(\gamma)} \prod_{i=1}^{m-1} \frac{\alpha\left(1-(1-\delta^i)a_i(\gamma)\right)}{(1-\delta^{i+1})a_i(\gamma)} < \infty. \tag{26}$$

Proof. It can be seen from (25) that if its denominator is finite then $p_{m,k}$ for $m \geq 1$ and $n \geq 0$ is a probability distribution and vica versa. □

The dependency on parameters λ and s are captured implicitly via $a_i(\gamma)$ in the above stability condition.

4.3   Marginal Stationary Distributions

The PGF of the marginal stationary distribution of the number of servers and the number of customers can be obtained from (25) by setting $z = 1$ and $u = 1$ in it, respectively.

Corollary 3. *The PGF of the marginal distribution of the number of servers in the stable controllable capacity queue is given as*

$$p(u,1) = \frac{\sum_{m=1}^{\infty} u^{m-1} \frac{1}{a_m(\gamma)} \prod_{i=1}^{m-1} \frac{\alpha\left(1-(1-\delta^i)a_i(\gamma)\right)}{(1-\delta^{i+1})a_i(\gamma)}}{\sum_{m=1}^{\infty} \frac{1}{a_m(\gamma)} \prod_{i=1}^{m-1} \frac{\alpha(1-(1-\delta^i)a_i(\gamma))}{(1-\delta^{i+1})a_i(\gamma)}}. \tag{27}$$

Corollary 4. *The PGF of the marginal distribution of the number of customers in the stable controllable capacity queue is given as*

$$p(1,z) = \frac{\sum_{m=1}^{\infty} \frac{a_m(z)}{a_m(\gamma)} \prod_{i=1}^{m-1} \frac{\alpha\left(1-(1-\delta^i)a_i(\gamma)\right)}{(1-\delta^{i+1})a_i(\gamma)}}{\sum_{m=1}^{\infty} \frac{1}{a_m(\gamma)} \prod_{i=1}^{m-1} \frac{\alpha(1-(1-\delta^i)a_i(\gamma))}{(1-\delta^{i+1})a_i(\gamma)}}. \tag{28}$$

5   System Measures

In this section we provide formulas for the most important system measures including the mean stationary number of servers, the mean stationary number of customers and the blocking probability.

5.1   Mean Stationary Number of the Servers and the Customers

The stationary number of the servers at the embedded epochs is given as

$$M = \lim_{n \to \infty} M(t^{d+}(n)).$$

Similarly the stationary number of the customers at the embedded epochs is given as

$$N = \lim_{n \to \infty} N(t^{d+}(n)).$$

The expression of the mean stationary number of servers and mean number of customers can be obtained from (25) by taking $\frac{\delta p(u,1)}{\delta u}\Big|_{u=1}$ and $\frac{\delta p(1,z)}{\delta z}\Big|_{z=1}$ of it, respectively.

Corollary 5. *The mean stationary number of servers in the stable controllable capacity queue is given as*

$$E[M] = \frac{\sum_{m=1}^{\infty}(m-1)\frac{1}{a_m(\gamma)}\prod_{i=1}^{m-1}\frac{\alpha\left(1-(1-\delta^i)a_i(\gamma)\right)}{(1-\delta^{i+1})a_i(\gamma)}}{\sum_{m=1}^{\infty}\frac{1}{a_m(\gamma)}\prod_{i=1}^{m-1}\frac{\alpha(1-(1-\delta^i)a_i(\gamma))}{(1-\delta^{i+1})a_i(\gamma)}}. \tag{29}$$

Corollary 6. *The mean stationary number of customers in the stable controllable capacity queue is given as*

$$E[N] = \frac{\sum_{m=1}^{\infty}\frac{a_m^{(1)}(1)}{a_m(\gamma)}\prod_{i=1}^{m-1}\frac{\alpha\left(1-(1-\delta^i)a_i(\gamma)\right)}{(1-\delta^{i+1})a_i(\gamma)}}{\sum_{m=1}^{\infty}\frac{1}{a_m(\gamma)}\prod_{i=1}^{m-1}\frac{\alpha(1-(1-\delta^i)a_i(\gamma))}{(1-\delta^{i+1})a_i(\gamma)}}. \tag{30}$$

5.2 Blocking Probability

We define the conditional probability b_m as the probability that the number of arrivals during the targeted service time $\geq m + 1$, given that the number of arrivals during the targeted service time $\geq m$, for $m \geq 1$.
The probability b_m can be given as

$$b_m = \frac{a_{m+1}^+}{a_m^+}. \tag{31}$$

Utilizing the independency of the arrival process of the change of the number of servers, the blocking probability p_b can be computed as

$$p_b = \sum_{m=1}^{\infty} b_m p_{m,m}. \tag{32}$$

Theorem 3. *The blocking probability in the stable controllable capacity queue can be expressed as*

$$p_b = \frac{\sum_{m=1}^{\infty}\frac{a_{m+1}^+}{a_m(\gamma)}\prod_{i=1}^{m-1}\frac{\alpha\left(1-(1-\delta^i)a_i(\gamma)\right)}{(1-\delta^{i+1})a_i(\gamma)}}{\sum_{m=1}^{\infty}\frac{1}{a_m(\gamma)}\prod_{i=1}^{m-1}\frac{\alpha(1-(1-\delta^i)a_i(\gamma))}{(1-\delta^{i+1})a_i(\gamma)}}. \tag{33}$$

Proof. The probabilities $p_{m,m}$ can be computed form (10) by taking $\frac{1}{m!}\left.\frac{d}{dz^m}\right|_{z=1}$ on it, which gives

$$p_{m,m} = \frac{a_m^+}{a_0}p_{m,0}, \quad m \geq 1. \tag{34}$$

The probabilities $p_{m,0}$ are computed from (15) by setting $z = 0$ leading to

$$p_{m,0} = \frac{\frac{a_0}{a_m(\gamma)}\prod_{i=1}^{m-1}\frac{\alpha\left(1-(1-\delta^i)a_i(\gamma)\right)}{(1-\delta^{i+1})a_i(\gamma)}}{\sum_{m=1}^{\infty}\frac{1}{a_m(\gamma)}\prod_{i=1}^{m-1}\frac{\alpha(1-(1-\delta^i)a_i(\gamma))}{(1-\delta^{i+1})a_i(\gamma)}}, \quad m \geq 1. \tag{35}$$

Applying (35), (34) and (31) in (32) results in

$$p_b = \frac{\sum_{m=1}^{\infty}\frac{a_{m+1}^+}{a_m^+}\frac{a_m^+}{a_0}\frac{a_0}{a_m(\gamma)}\prod_{i=1}^{m-1}\frac{\alpha\left(1-(1-\delta^i)a_i(\gamma)\right)}{(1-\delta^{i+1})a_i(\gamma)}}{\sum_{m=1}^{\infty}\frac{1}{a_m(\gamma)}\prod_{i=1}^{m-1}\frac{\alpha(1-(1-\delta^i)a_i(\gamma))}{(1-\delta^{i+1})a_i(\gamma)}}. \tag{36}$$

The statement of the theorem comes by rearranging (36). $\qquad\square$

6 Final Remarks

This work is intended to be extended in several directions including

- establishing cost model and performing optimization,
- generalizing the model to vacation model.

It is a topic of future research to analyze the loss-free counterpart of the considered model.

References

1. Malak, D., Dhillon, H.S., Andrews, J.G.: Optimizing data aggregation for uplink machine-to-machine communication networks. IEEE Trans. Commun. **64**(3), 1274–1290 (2016)
2. Wang, F.-Y.: Parallel control and management for intelligent transportation systems: Concepts, architectures, and applications. IEEE Trans. Intell. Transp. Syst. **11**(3), 630–638 (2010)
3. Saffer, Z., Telek, M.: Analysis of globally gated Markovian limited cyclic polling model and its application to uplink traffic in the IEEE 802.16 network. J. Ind. Manag. Optim. (JIMO) **7**(3), 677–697 (2011)
4. Bruneel, H., Walraevens, J., Claeys, D., Wittevrongel, S.: Analysis of a discrete-time queue with geometrically distributed service capacities. In: Al-Begain, K., Fiems, D., Vincent, J.-M. (eds.) ASMTA 2012. LNCS, vol. 7314, pp. 121–135. Springer, Heidelberg (2012). https://doi.org/10.1007/978-3-642-30782-9_9
5. Conway, R.W., Maxwell, W.L.: A queueing model with state dependent service rate. J. Ind. Eng. **12**, 132–136 (1961)
6. Lia, H., Yang, T.: Queues with a variable number of servers. EJOR **124**(3), 615–628 (2000)
7. Mazalov, V., Gurtov, A.: Queuing system with on-demand number of servers. Math. Appl. 40(2) **15/56**, 1–12 (2012)

Delay Analysis of a Queue with General Service Demands and Correlated Service Capacities

Michiel De Muynck$^{(\boxtimes)}$, Herwig Bruneel, and Sabine Wittevrongel

Department TELIN, Ghent University,
Sint-Pietersnieuwstraat 41, 9000 Ghent, Belgium
{MichielR.DeMuynck,Herwig.Bruneel,Sabine.Wittevrongel}@UGent.be

Abstract. We present the study of a non-classical discrete-time queueing model in which the customers each request a variable amount of service, called their "service demand", from a server which is able to execute a variable amount of work, called its "service capacity", during each time slot. We assume that the numbers of arrivals in consecutive time slots and the service demands of consecutive customers form two independent and identically distributed sequences. However, we allow the service capacities in consecutive time slots to be correlated according to a discrete-batch Markovian process. We study this model analytically and obtain an expression for the probability generating function of the delay of an arbitrary customer in steady state. The results are illustrated with several numerical examples.

Keywords: Discrete-time queueing theory · Service demands
Correlated service capacities · Discrete-batch Markovian service process

1 Introduction

In classical queueing theory, queueing phenomena where customers require varying amounts of work from the server(s) are often modeled using the concept of "service time", where the service time of a customer is the amount of time that a server needs to fully process that customer. It is then commonly assumed that the service times of the consecutive customers are independent from each other.

However, in many queueing phenomena this assumption may not hold. Indeed, the service time of a customer is usually determined by two underlying quantities: the amount of work that the customer requires from the server, which we refer to as the "service demand" of that customer, and the speed with which the server can process this work. The assumption that service times form an independent and identically distributed (i.i.d.) sequence generally breaks down into assuming that both these service demands and the service speeds are uncorrelated from customer to customer.

The first of these two assumptions, i.e., that the service demands of subsequent customers are independent of each other, is valid for many queueing

© Springer International Publishing AG, part of Springer Nature 2018
Y. Takahashi et al. (Eds.): QTNA 2018, LNCS 10932, pp. 64–85, 2018.
https://doi.org/10.1007/978-3-319-93736-6_5

phenomena. For example, in most brick-and-mortar stores there is little correlation between the length of the shopping list of one customer entering that store and that of the next. However, the second assumption, i.e., that the speed with which subsequent customers are served is also uncorrelated, is often not valid. If the cashier at a register is working slowly, it is likely that the next person served at that register will also receive slower service.

Other examples of queueing phenomena where there is little correlation in the service demands while the speed with which a server processes two consecutive customers is usually correlated are web services where the available processing power fluctuates due to background processes or shared hosting, wireless communication channels where the available bandwidth fluctuates over time due to interference and varying conditions (see e.g. [1,2]), manufacturing facilities where the production capacity varies over time due to maintenance and repairs (see e.g. [3,4]), etc.

While there is some research in the scientific literature about continuous-time queueing models with both variable service demands and variable rates of service (see e.g. [5–7]), discrete-time queueing models with variable service demands and capacities have received comparatively little attention. In this paper, we will model these queueing phenomena using a discrete-time queueing model, meaning time is divided into evenly spaced time slots. The number of work units that the server can execute in a given slot is referred to as the "service capacity" of the server during that time slot, which we assume to be a non-negative integer. The service demands of all customers are assumed to be (positive) integer numbers of work units as well. We assume that available service capacity is never wasted, i.e., in one time slot multiple customers may be served if there is enough service capacity to do so. Similarly, if the available service capacity is not enough to fully serve the customer in service, the server can execute some of the work units of the service demand of that customer so that these work units do not need to be executed again in a later slot. For more details about the queueing model, see Sect. 2.

There have been multiple papers in the literature studying this non-classical queueing model, with varying restrictions on the distribution of the service capacities in a given time slot. In [8,9], results were obtained for geometrically distributed service capacities, while in [10,11], the service capacity was assumed to be constant, i.e., the same in all time slots. In [12], the distribution of the service capacities was assumed to have finite support, while in [13,14] the service capacities were assumed to follow a phase-type distribution. However, in all of these studies, the service capacities were assumed to be independent from slot to slot. In this paper, we allow the service capacities to be correlated by assuming that they follow a discrete-batch Markovian process, which is most commonly used to describe the number of arriving customers in consecutive time slots, and is then usually called the discrete-batch Markovian arrival process (D-BMAP), see e.g. [15].

This paper is organized as follows. In Sect. 2 we describe the mathematical queueing model under study, as well as the notation used in its analysis. Then, in

Sect. 3, we study a special case of the queueing model where the service demands are assumed to be equal to 1 work unit. The results obtained for this special case are then used in Sect. 4, where we obtain results for the probability generating function (pgf) of the delay of an arbitrary customer in steady state, and in Sect. 5, we describe how to evaluate the mean and other moments of this customer delay. In Sect. 6, we illustrate the model using several numerical examples, and finally Sect. 7 gives a brief conclusion of the paper. Appendix A outlines how models with phase-type vacations can be studied, as they are a special case of the model studied in this paper.

2   Queueing Model and Notation

In this section, we give a formal description of the queueing model studied in this paper, as well as the mathematical notation used in its analysis. The studied queueing model is a discrete-time model, i.e., time is divided into discrete fixed-length intervals, referred to as (time) slots. During each time slot, a random number of customers arrive to the queue. The number of customers arriving in slot k is denoted by A_k. The sequence $A_1, A_2, \ldots$ is assumed to be an i.i.d. sequence with common pgf $A(z)$. The mean number of customers arriving per slot is denoted by $\lambda \triangleq A'(1)$.

Each customer requires a certain amount of service from the server referred to as the *service demand* of that customer. The service demand of customer k is assumed to consist of an integer number of *work units* and is denoted by S_k. The sequence $S_1, S_2, \ldots$ forms an i.i.d. sequence with pgf $S(z)$ and mean $\tau \triangleq S'(1)$.

There is one server in the queueing system. During each time slot it is able to execute an integer number of work units, referred to as the *service capacity* of the server during that time slot. The service capacity during slot k is denoted by R_k. These R_k work units of service capacity are always executed in a work-conserving manner, i.e., if R_k is larger than the (remaining) service demand of the customer in service, then the next customer immediately begins service (during the same slot). The only exception to this is that arriving customers can never receive any service during their arrival slot, i.e., the system is a late-arrival system with delayed access (LAS-DA). If R_k is less than the remaining service demand of the customer in service, then the remaining service demand of that customer decreases by R_k and the service of that customer simply continues in the next slot.

The service capacities are not uncorrelated from slot to slot. Instead, they depend on a background Markov chain with m states. The state of that Markov chain during slot k is denoted by T_k, with $0 \leq T_k \leq m - 1$. We assume that the sequence $T_1, T_2, \ldots$ forms an irreducible and aperiodic discrete-time Markov chain. Its transition matrix is denoted by $\mathbf{R}(1)$ and the equilibrium distribution of this Markov chain is denoted by the row-vector $\boldsymbol{\pi}$. It is well-known that $\boldsymbol{\pi}$ is the unique probability row-vector satisfying

$$\boldsymbol{\pi}\mathbf{R}(1) = \boldsymbol{\pi}. \tag{1}$$

The service capacity R_k during slot k is completely determined by T_k and T_{k+1}. In particular, the service process is entirely defined by the probabilities $P(R_k = n, T_{k+1} = j|T_k = i)$, which are the same for all k. We can gather all these probabilities in a single matrix generating function $\mathbf{R}(z)$, whose (i,j)th entry is given by

$$[\mathbf{R}(z)]_{i,j} \triangleq \sum_{n=0}^{\infty} P(R_k = n, T_{k+1} = j|T_k = i)z^n.$$

Note that this definition is consistent with the earlier definition of the transition matrix $\mathbf{R}(1)$ of the background Markov chain. The mean service capacity in an arbitrary slot in steady state, which we denote by μ, can be found as $\boldsymbol{\pi}\mathbf{R}'(1)\mathbf{1}^T$, where $\mathbf{1}^T$ denotes the column vector $[1\ 1\ ...\ 1]^T$ of the appropriate dimension depending on the context (here m). We assume that the mean service capacity μ in steady state is larger than the mean total service demand $\lambda\tau$ of all customers arriving in an arbitrary slot in steady state, i.e., we assume that the load $\rho \triangleq \lambda\tau/\mu$ is smaller than 1. We also assume that all elements of $[\mathbf{R}(z)]_{i,j}$ are polynomial functions, which is the case if there is a maximum service capacity that the server cannot exceed.

The number of customers in the system (i.e., the "system content") at the start of slot k is denoted by B_k. The steady-state pgf of B_k as $k \to \infty$ is denoted $B(z)$. To account for the correlation between the system content and the state of the background Markov chain, we also define the partial pgfs

$$[\mathbf{B}(z)]_j = \lim_{k\to\infty} E[z^{B_k}I(T_k = j)], \tag{2}$$

where $I(...)$ is the indicator function. Together, these m partial pgfs form the row vector $\mathbf{B}(z)$. Note that the sum of all elements of $\mathbf{B}(z)$ is equal to the pgf of the system content at the start of an arbitrary slot in steady state, which is denoted as $B(z)$.

3   Service Demands of 1 Work Unit

In this section, we study a special case of the queueing model specified in the previous section. Namely, we study the case where all service demands are equal to 1 work unit, i.e., $S(z) = z$. The results of this analysis will turn out to be useful in the subsequent sections. Additionally, this special case is noteworthy in its own right, since it is equivalent to the discrete-time $Geo^X/DBMSP/1$ classical queueing model. In this special case, the customers are served according to a discrete-time batch Markovian service process (DBMSP), which is analogous to the well-known discrete-time batch Markovian arrival process (DBMAP, see e.g. [15]). A version of the $Geo^X/DBMSP/1$ queuing model with finite buffer space was studied in [16]. To the best of our knowledge, no results for the pgf of the steady-state system content of $Geo^X/DBMSP/1$ queues with infinite buffer space have been reported in the literature.

In this special case, there is a very simple system equation relating B_{k+1} to B_k, namely

$$B_{k+1} = (B_k - R_k)^+ + A_k, \tag{3}$$

where $(X)^+$ denotes $\max(X, 0)$. Indeed, each work unit of service capacity corresponds to exactly one customer leaving the system, if there is one. To emphasize that the results obtained in this section are only valid for the special case of service demands of 1 work unit, we will in this section use the notation $\tilde{}$ on all variables relating to the system content, e.g., B_k and $\mathbf{B}(z)$ will be denoted $\tilde{B}_k$ and $\tilde{\mathbf{B}}(z)$ respectively, and A_k and $A(z)$ are denoted $\tilde{A}_k$ and $\tilde{A}(z)$. For example, the Eq. (3) above is rewritten as

$$\tilde{B}_{k+1} = (\tilde{B}_k - R_k)^+ + A_k. \tag{4}$$

Similarly, we rewrite Eq. (2) as

$$
\begin{aligned}
[\tilde{\mathbf{B}}(z)]_j &= \lim_{k \to \infty} E[z^{\tilde{B}_k} I(T_k = j)] \\
&= \lim_{k \to \infty} \sum_{n=0}^{\infty} \sum_{i=1}^{m} E[z^{\tilde{B}_k} I(T_k = j, T_{k-1} = i, \tilde{B}_{k-1} = n)] \\
&= \lim_{k \to \infty} \sum_{n=0}^{\infty} \sum_{i=1}^{m} E[z^{\tilde{B}_k} I(T_k = j) | T_{k-1} = i, \tilde{B}_{k-1} = n] \\
&\quad \cdot P(T_{k-1} = i, \tilde{B}_{k-1} = n).
\end{aligned}
\tag{5}
$$

The factor $E[z^{\tilde{B}_k} I(T_k = j) | T_{k-1} = i, \tilde{B}_{k-1} = n]$ in the above equation can be rewritten using Eq. (4) as

$$
\begin{aligned}
E[z^{\tilde{B}_k} I(T_k = j) | T_{k-1} = i, \tilde{B}_{k-1} = n] \\
= E[z^{(n-R_{k-1})^+ + \tilde{A}_{k-1}} I(T_k = j) | T_{k-1} = i, \tilde{B}_{k-1} = n] \\
= \tilde{A}(z) E[z^{(n-R_{k-1})^+} I(T_k = j) | T_{k-1} = i],
\end{aligned}
\tag{6}
$$

due to the independence of $\tilde{A}_{k-1}$, $\tilde{B}_{k-1}$, and (T_k, R_{k-1}) when given T_{k-1}.

To continue, we use the inversion formula for probability generating functions, which states that for a random variable X with pgf $X(z)$ with radius of convergence $\mathcal{R}_X$,

$$P(X = n) = \frac{1}{2\pi i} \oint_L \frac{X(\zeta)}{\zeta^{n+1}} \, d\zeta, \tag{7}$$

where i indicates the imaginary unit and L is any counterclockwise contour around the origin where $\forall \zeta \in L : |\zeta| < \mathcal{R}_X$. Using this, we can now rewrite (6) further as

$$E[z^{\tilde{B}_k} I(T_k = j)|T_{k-1} = i, \tilde{B}_{k-1} = n]$$

$$= \tilde{A}(z) \left(\sum_{l=0}^{n-1} z^{n-l} P(T_k = j, R_{k-1} = l|T_{k-1} = i) \right.$$

$$\left. + \sum_{l=n}^{\infty} P(R_{k-1} = l, T_k = j|T_{k-1} = i) \right)$$

$$= \frac{\tilde{A}(z)}{2\pi\imath} \left(\sum_{l=0}^{n-1} z^{n-l} \oint_L \frac{[\mathbf{R}(\zeta)]_{i,j}}{\zeta^{l+1}} \, \mathrm{d}\zeta + \sum_{l=n}^{\infty} \oint_L \frac{[\mathbf{R}(\zeta)]_{i,j}}{\zeta^{l+1}} \, \mathrm{d}\zeta \right), \qquad (8)$$

where L is any counterclockwise contour around the origin where $\forall \zeta \in L : |\zeta| < \mathcal{R}_\mathbf{R}$, in which $\mathcal{R}_\mathbf{R}$ denotes the minimum of the radii of convergence of all elements of $\mathbf{R}(z)$.

We may now interchange the order of the infinite summation over l and the integration over L in (8) (i.e., the infinite summation of the contour integral is equal to the contour integral of the infinite summation) if the infinite series in the resulting integrand converges uniformly. This is the case if $\forall \zeta \in L : |1/\zeta| < 1$. If we restrict L to be any contour where that is the case, we may write

$$E[z^{\tilde{B}_k} I(T_k = j)|T_{k-1} = i, \tilde{B}_{k-1} = n]$$

$$= \frac{\tilde{A}(z)}{2\pi\imath} \left(z \oint_L \frac{[\mathbf{R}(\zeta)]_{i,j}}{\zeta^n} \frac{(z\zeta)^n - 1}{z\zeta - 1} \, \mathrm{d}\zeta + \oint_L \frac{[\mathbf{R}(\zeta)]_{i,j}}{\zeta^n} \frac{1}{\zeta - 1} \, \mathrm{d}\zeta \right)$$

$$= \frac{\tilde{A}(z)}{2\pi\imath} \oint_L [\mathbf{R}(\zeta)]_{i,j} \frac{z^{n+1}(\zeta - 1) + \zeta^{-n}(z - 1)}{(z\zeta - 1)(\zeta - 1)} \, \mathrm{d}\zeta.$$

Next, we substitute this back into (5). In the resulting equation, we may swap the summation over n with the contour integral over L if $|z| < \mathcal{R}_{\tilde{\mathbf{B}}}$ and $\forall \zeta \in L : |1/\zeta| < \mathcal{R}_{\tilde{\mathbf{B}}}$. If we restrict L further so that this is also the case, we may write

$$[\tilde{\mathbf{B}}(z)]_j = \sum_{i=1}^{m} \frac{\tilde{A}(z)}{2\pi\imath} \oint_L [\mathbf{R}(\zeta)]_{i,j} \frac{z[\tilde{\mathbf{B}}(z)]_i(\zeta - 1) + [\tilde{\mathbf{B}}(1/\zeta)]_i(z - 1)}{(z\zeta - 1)(\zeta - 1)} \, \mathrm{d}\zeta$$

$$= \frac{\tilde{A}(z)}{2\pi\imath} \oint_L \left[\frac{z\tilde{\mathbf{B}}(z)}{z\zeta - 1} \mathbf{R}(\zeta) \right]_j \, \mathrm{d}\zeta + \frac{\tilde{A}(z)(z - 1)}{2\pi\imath} \oint_L \left[\frac{\tilde{\mathbf{B}}(1/\zeta)\mathbf{R}(\zeta)}{(z\zeta - 1)(\zeta - 1)} \right]_j \, \mathrm{d}\zeta.$$

$$(9)$$

The last step is only valid of $1/z$ is not on the contour L, so we further restrict L so that $1/z$ does not lie on it. It is, however, not specified whether this point $1/z$, which is a pole of both integrands in (9), lies inside or outside L. With the restrictions we have placed on L so far, i.e., that that L must be a counter-clockwise contour around the origin where $\forall \zeta \in L : 1/\mathcal{R}_{\tilde{\mathbf{B}}} < 1 < |\zeta| < \mathcal{R}_\mathbf{R}$ and $\zeta \neq 1/z$, it is possible that some of the contours that satisfy all these conditions (called "valid" contours) have $1/z$ as an interior point while others don't. If $|z| < 1$, we can always guarantee that a valid contour exists that doesn't encircle $1/z$, while if $|z| \geq 1$, any valid contour must encircle $1/z$. In the sequel we will show the detailed derivation of $\tilde{B}(z)$ for the case where $|z| < 1$, choosing L to be a valid contour that doesn't encircle $1/z$. It is easily verified that if $|z| \geq 1$, a similar derivation leads to the same expression for $\tilde{B}(z)$. This derivation is by itself however not very interesting, so it is omitted here.

In the case where $|z| < 1$ and where the contour L does not encircle $1/z$, the integrand in the first term on the right-hand side of (9) does not have any poles inside L, so

$$\frac{\tilde{A}(z)}{2\pi\iota} \oint_L \left[\frac{z\tilde{B}(z)}{z\zeta - 1} \mathbf{R}(\zeta) \right]_j d\zeta = 0. \tag{10}$$

The integrand in the second term, however, does have poles inside L. These poles are the poles of the elements of $\tilde{\mathbf{B}}(1/\zeta)$, which we don't know yet. We therefore perform the substitution $\zeta = 1/\xi$ (which yields a factor $-1/\xi^2$ in the integrand) but still integrate in the counterclockwise sense (which yields an extra factor -1). Denoting the new counterclockwise integration path as L', we rewrite the second term of the right-hand side of (9) as

$$\frac{A(z)(z-1)}{2\pi\iota} \oint_L \left[\frac{\tilde{\mathbf{B}}(1/\zeta)\mathbf{R}(\zeta)}{(z\zeta-1)(\zeta-1)} \right]_j d\zeta = \frac{A(z)(z-1)}{2\pi\iota} \oint_{L'} \left[\frac{\tilde{\mathbf{B}}(\xi)\mathbf{R}(1/\xi)}{(z-\xi)(1-\xi)} \right]_j d\xi \tag{11}$$

The poles of the new integrand inside L' are $\xi = z$ and the poles of the elements of $\mathbf{R}(1/\xi)$ on the jth column. We denote the set of these latter poles as $[\mathcal{S}_\mathbf{R}]_j$. If $z \notin [\mathcal{S}_\mathbf{R}]_j$, the residue of the integrand at $\xi = z$ is given by

$$\operatorname*{Res}_{\xi \to z} \left[\frac{\tilde{\mathbf{B}}(\xi)\mathbf{R}(1/\xi)}{(z-\xi)(1-\xi)} \right]_j = \frac{[\tilde{\mathbf{B}}(z)\mathbf{R}(1/z)]_j}{z-1}$$

and the residue at a pole $\xi = \xi^*$ with multiplicity m_{ξ^*} is given by

$$\operatorname*{Res}_{\xi \to \xi^*} \left[\frac{\tilde{\mathbf{B}}(\xi)\mathbf{R}(1/\xi)}{(z-\xi)(1-\xi)} \right]_j = \frac{1}{(m_{\xi^*}-1)!} \lim_{\xi \to \xi^*} \frac{d^{m_{\xi^*}-1}}{d\xi^{m_{\xi^*}-1}} (\xi - \xi^*)^{m_{\xi^*}} \frac{[\tilde{\mathbf{B}}(\xi)\mathbf{R}(1/\xi)]_j}{(z-\xi)(1-\xi)}$$

$$= \sum_{n=1}^{m_{\xi^*}} \frac{\tilde{c}_{\xi^*,n,j}}{(z-\xi^*)^n}, \tag{12}$$

for some constants $\tilde{c}_{\xi^*,n,j}$.

Substituting this and (10) back into (9) and using Cauchy's residue theorem, we get for $z \notin [\mathcal{S}_\mathbf{R}]_j$,

$$[\tilde{\mathbf{B}}(z)]_j = \tilde{A}(z)[\tilde{\mathbf{B}}(z)\mathbf{R}(1/z)]_j + \tilde{A}(z)(z-1)[\tilde{\mathbf{C}}(z)]_j. \tag{13}$$

Here $\tilde{\mathbf{C}}(z)$ is the row vector defined as

$$[\tilde{\mathbf{C}}(z)]_j = \sum_{\xi^* \in [\mathcal{S}_\mathbf{R}]_j} \sum_{n=1}^{m_{\xi^*}} \frac{\tilde{c}_{\xi^*,n,j}}{(z - \xi^*)^n} = \frac{f_j(z)}{\prod_{\xi^* \in [\mathcal{S}_\mathbf{R}]_j} (z - \xi^*)^{m_{\xi^*}}},$$

for some polynomial $f_j(z)$ with degree equal to $m_{\mathbf{R},j} - 1$, where $m_{\mathbf{R},j}$ is the sum of all the multiplicities of all the poles in $[\mathcal{S}_\mathbf{R}]_j$. If $m_{\mathbf{R},j} = 0$ then $f_j(z) = 0$. If we rewrite the polynomial $f_j(z)$ as

$$f_j(z) = \tilde{\mathbf{C}}_j \mathbf{Z}_j(z),$$

where

$$\tilde{\mathbf{C}}_j = \begin{bmatrix} \tilde{c}_{j,0} \ \tilde{c}_{j,1} \ \tilde{c}_{j,2} \ \cdots \ \tilde{c}_{j,m_{\mathbf{R},j}-1} \end{bmatrix}, \qquad \mathbf{Z}_j(z) = \begin{bmatrix} 1 \ z \ z^2 \ \ldots \ z^{m_{\mathbf{R},j}-1} \end{bmatrix}^T,$$

then we can rewrite $\tilde{\mathbf{C}}(z)$ as

$$\tilde{\mathbf{C}}(z) = \tilde{\mathbf{C}}\mathbf{Z}(z)\mathbf{P}(z)^{-1}, \tag{14}$$

where $\tilde{\mathbf{C}}$ denotes the row vector $\begin{bmatrix} \tilde{\mathbf{C}}_1 \ \tilde{\mathbf{C}}_2 \ldots \tilde{\mathbf{C}}_m \end{bmatrix}$, $\mathbf{Z}(z)$ denotes the block-diagonal matrix $\mathrm{diag}(\mathbf{Z}_1, \mathbf{Z}_2, \ldots, \mathbf{Z}_m)$ and $\mathbf{P}(z)$ denotes the diagonal matrix

$$\mathbf{P}(z) = \mathrm{diag}\left(\prod_{\xi^* \in [\mathcal{S}_\mathbf{R}]_1} (z - \xi^*)^{m_{\xi^*}}, \prod_{\xi^* \in [\mathcal{S}_\mathbf{R}]_2} (z - \xi^*)^{m_{\xi^*}}, \ldots, \prod_{\xi^* \in [\mathcal{S}_\mathbf{R}]_m} (z - \xi^*)^{m_{\xi^*}} \right).$$

Combining the result of $[\tilde{\mathbf{B}}(z)]_j$ from Eq. (13) for each of the m values of j into one matrix equation, we find

$$\tilde{\mathbf{B}}(z) = \tilde{A}(z)\tilde{\mathbf{B}}(z)\mathbf{R}(1/z) + \tilde{A}(z)(z - 1)\tilde{\mathbf{C}}(z).$$

Solving for $\tilde{\mathbf{B}}(z)$ and using (14), we get

$$\tilde{\mathbf{B}}(z) = \tilde{A}(z)(z - 1)\tilde{\mathbf{C}}\mathbf{Z}(z)\mathbf{P}(z)^{-1}\big(\mathbf{I} - \tilde{A}(z)\mathbf{R}(1/z)\big)^{-1}, \tag{15}$$

or equivalently,

$$\tilde{\mathbf{B}}(z) = \tilde{A}(z)(z - 1)\tilde{\mathbf{C}}\mathbf{Z}(z)\frac{\mathrm{adj}(\mathbf{P}(z) - \tilde{A}(z)\mathbf{R}(1/z)\mathbf{P}(z))}{\det(\mathbf{P}(z) - \tilde{A}(z)\mathbf{R}(1/z)\mathbf{P}(z))}. \tag{16}$$

All that remains now is the determination of the $m_\mathbf{R} \triangleq m_{\mathbf{R},1} + \ldots + m_{\mathbf{R},m}$ unknown constants in $\tilde{\mathbf{C}}$. For this, we will use the following theorem:

Theorem 1 (Matrix version of Rouché's theorem). *Let $\mathbf{F}(z)$ and $\mathbf{G}(z)$ be two analytic $m \times m$ matrix-valued functions inside and on a Cauchy contour Γ. If $\mathbf{F}(z)$ is not singular on Γ and*

$$||\mathbf{F}^{-1}(z)\mathbf{G}(z)|| < 1 \ \forall \ z \in \Gamma \quad or \quad ||\mathbf{G}(z)\mathbf{F}^{-1}(z)|| < 1 \ \forall \ z \in \Gamma,$$

where $|| \cdot ||$ denotes the operator norm, then $\mathbf{F}(z) + \mathbf{G}(z)$ is not singular on Γ, and $\mathbf{F}(z)$ has the same number of singularities (i.e., roots of $\det \mathbf{F}(z)$, counting with multiplicity) as $\mathbf{F}(z) + \mathbf{G}(z)$ inside Γ.

Proof. See Sect. 5.3 of [17].

Using this theorem with $\mathbf{F}(z) = \mathbf{P}(z)$ and $\mathbf{G}(z) = -\tilde{A}(z)\mathbf{R}(1/z)\mathbf{P}(z)$ and Γ a circle with radius $1 + \epsilon$ around the origin, it can be shown that $\det(\mathbf{P}(z) - \tilde{A}(z)\mathbf{R}(1/z)\mathbf{P}(z))$ has the same number of zeros inside or on the unit circle as $\det(\mathbf{P}(z))$ (counting with multiplicity). It is easy to see that this number of zeros is equal to $m_{\mathbf{R}}$, since all the zeros of $\det(\mathbf{P}(z))$ lie inside or on the unit circle. We denote the set of the zeros of $\det(\mathbf{P}(z) - \tilde{A}(z)\mathbf{R}(1/z)\mathbf{P}(z))$ as $\mathcal{S}_{\tilde{\mathbf{B}}}$. It is easy to see that $1 \in \mathcal{S}_{\tilde{\mathbf{B}}}$, because $\mathbf{R}(1)$ is a stochastic matrix, so $\mathbf{I} - \mathbf{R}(1)$ is singular.

Since each element of $\tilde{\mathbf{B}}(z)$ is a partial pgf, it must remain bounded inside and on the unit circle. Therefore, the numerator of (16) must become $\mathbf{0}$ with the same multiplicity as well, where $\mathbf{0}$ is the row vector $[0\ 0\ ...\ 0]$ of the appropriate dimension (here m). This implies that for each $z^* \in \mathcal{S}_{\tilde{\mathbf{B}}}$ with $z^* \neq 1$,

$$\tilde{\mathbf{C}}\mathbf{Z}(z^*)\,\mathrm{adj}\,\left(\mathbf{P}(z^*) - \tilde{A}(z^*)\mathbf{R}(1/z^*)\mathbf{P}(z^*)\right) = \mathbf{0}. \tag{17}$$

If z^* is a zero of multiplicity $n > 1$, then this also implies that the 1st, 2nd, ..., $(n-1)$th derivatives of the numerator of (16) must become $\mathbf{0}$. Since these $(n-1)$th order derivatives do not simplify further in the general case, they need to be studied on an ad hoc basis. Luckily, in most practical examples the zeros $\mathcal{S}_{\tilde{\mathbf{B}}}$ do have multiplicity 1, so in the remainder of this paper we will assume that the zeros $\mathcal{S}_{\tilde{\mathbf{B}}}$ are indeed distinct, i.e., they have multiplicity 1. See [18, Appendix A] for a more detailed explanation on a similar problem.

For each $z^* \in \mathcal{S}_{\tilde{\mathbf{B}}}$, the expression (17) gives us m linear equations for the unknown constants in the matrix $\tilde{\mathbf{C}}$. Unfortunately, it turns out that these m equations are all linearly dependent, so each z^* only contributes 1 linearly independent equation. Since $z^* = 1$ does not contribute a linear equation (the numerator of (16) vanishes at $z = 1$ regardless of the value of $\tilde{\mathbf{C}}$), we now have $|\mathcal{S}_{\tilde{\mathbf{B}}}| - 1$ independent equations. We can obtain one final equation by considering the normalization condition, i.e., $\tilde{B}(1) = \tilde{\mathbf{B}}(1)\mathbf{1}^T = 1$. Using l'Hôpital's rule and Jacobi's formula, we obtain

$$\tilde{\mathbf{C}}\mathbf{Z}(1)\,\mathrm{adj}\,\mathbf{P}(1)\,\mathrm{adj}\,\left(\mathbf{I} - \mathbf{R}(1)\right)\mathbf{1}^T = \tag{18}$$
$$\mathrm{tr}\left(\,\mathrm{adj}\,\mathbf{P}(1)\,\mathrm{adj}(\mathbf{I} - \mathbf{R}(1))\left(\mathbf{P}'(1) - \tilde{A}'(1)\mathbf{R}(1)\mathbf{P}(1) + \mathbf{R}'(1)\mathbf{P}(1) - \mathbf{R}(1)\mathbf{P}'(1)\right)\right).$$

This equation in addition to one of the equations from (17) for each of the $z^* \in \mathcal{S}_{\tilde{\mathbf{B}}}, z^* \neq 1$ together form a set of $|\mathcal{S}_{\tilde{\mathbf{B}}}|$ linearly independent equations for the $m_{\mathbf{R}} = |\mathcal{S}_{\tilde{\mathbf{B}}}|$ unknown constants in $\tilde{\mathbf{C}}$, from which $\tilde{\mathbf{C}}$ can be determined. Then $\tilde{\mathbf{B}}(z)$ follows immediately from (16), and $\tilde{B}(z)$ follows immediately from that by $\tilde{B}(z) = \tilde{\mathbf{B}}(z)\mathbf{1}^T$.

4   Customer Delay

In this section, we will derive an expression for the pgf $D(z)$ of the delay D_C experienced by an arbitrary customer C in steady state, under a FCFS scheduling discipline. We will do this in two steps. Informally, we will first measure the delay "in work units", and then in time slots. More formally, we will first study the minimum number V_C of work units that need to be executed by the server before the arbitrary customer C can leave the system, i.e., the sum of all (remaining) service demands of all customers in the queue minus the service demands of the customers behind customer C in the queue at the beginning of the slot after the arrival slot of customer C. We refer to this quantity V_C as the *unfinished work observed by customer* C. In a second step, we will use this distribution to derive an expression for the pgf of the delay D_C experienced by customer C.

4.1   Unfinished Work Observed by a Customer

As indicated, we will start by deriving an expression for the pgf $V(z)$ of the unfinished work V_C observed by an arbitrary customer C in steady state. Since in the later sections, the dependency between V_C and the states of the background Markov chain will also be important, for $j = 1, ..., m$ we define the partial pgf

$$[\mathbf{V}(z)]_j = E[z^{V_C} I(T_{K_C+1} = j)],$$

where slot K_C denotes the arrival slot of customer C. Together, these m partial pgfs form the row vector $\mathbf{V}(z)$, from which the pgf $V(z)$ can easily be found as $V(z) = \mathbf{V}(z)\mathbf{1}^T$.

   Consider first a specific customer C, not necessarily in steady state. The unfinished work V_C observed by the Cth customer is the sum of three independent random variables:

1. The number of work units present in the system at the beginning of the arrival slot of customer C that are still not executed at the start of the next slot. This first part is given by $(U_{K_C} - R_{K_C})^+$, where U_k denotes the total unfinished work at the start of a slot k.
2. The total service demand of all customers that arrive during slot K_C but "before" customer C (i.e., having service priority over customer C). If we denote the number of those customers as F_C, then the second part of the unfinished work V_C is given by $\sum_{n=1}^{F_C} S_{C-n}$.
3. The service demand of customer C itself. Per definition this is given by S_C.

In summary we have that

$$V_C = (U_{K_C} - R_{K_C})^+ + \sum_{n=1}^{F_C} S_{C-n} + S_C.$$

We therefore find that

$$E[z^{V_C} I(T_{K_C+1} = j)] = E[z^{(U_{K_C} - R_{K_C})^+ + \sum_{n=1}^{F_C} S_{C-n} + S_C} I(T_{K_C+1} = j)]$$

$$= E[z^{(U_{K_C} - R_{K_C})^+} I(T_{K_C+1} = j)]\, F(S(z))\, S(z), \quad (19)$$

where $F(z)$ is the pgf of F_C, which is the same for all C. In the last step we used the fact that the service times of the customers are i.i.d. and that the number of customers arriving in slot K_C and their service times are independent of U_{K_C}, R_{K_C} and T_{K_C+1}. It is well known (see e.g. [19]) that for a system with uncorrelated arrivals, $F(z)$ is given by

$$F(z) = \frac{A(z) - 1}{A'(1)(z - 1)}. \tag{20}$$

The first factor in (19) can be calculated from the system equation for U_k, i.e.,

$$U_{k+1} = (U_k - R_k)^+ + \sum_{n=1}^{A_k} S_{k,n},$$

where $S_{k,n}$ denotes the service demand of the nth customer arriving in slot k. This equation implies that

$$E[z^{U_{k+1}} I(T_{k+1} = j)] = E[z^{(U_k - R_k)^+} I(T_{k+1} = j)] A(S(z)),$$

where we again used the fact that both the arrival process and the service demands are uncorrelated. Substituting this result into (19) and considering an arbitrary customer C in steady state, we find

$$[\mathbf{V}(z)]_j = \frac{[\mathbf{U}(z)]_j}{A(S(z))} \frac{A(S(z)) - 1}{A'(1)(S(z) - 1)} S(z),$$

where $[\mathbf{U}(z)]_j = \lim_{k \to \infty} E[z^{U_k} I(T_k = j)]$.

An expression for the partial pgfs $[\mathbf{U}(z)]_j$ can be found using the techniques from Sect. 3. Indeed, the total unfinished work in a system where the arrivals and service demands have pgfs $A(z)$ and $S(z)$ respectively has the same distribution as the number of customers in a system where the arrivals have pgf $A(S(z))$ and all service demands are equal to 1 work unit. The special case of service demands of one work unit is exactly what was studied in Sect. 3. The row vector $\mathbf{U}(z)$ can therefore be found from the expression (15) for $\tilde{B}(z)$ with the substitution $\tilde{A}(z) = A(S(z))$. This leads to

$$\mathbf{V}(z) = \frac{S(z)}{A'(1)} \frac{A(S(z)) - 1}{S(z) - 1} (z - 1) \mathbf{C} \mathbf{Z}(z) \mathbf{P}(z)^{-1} (\mathbf{I} - A(S(z)) \mathbf{R}(1/z))^{-1}, \tag{21}$$

where $\mathbf{C}$ can be determined in the same way as $\tilde{\mathbf{C}}$ in Sect. 3, and $\mathbf{Z}(z)$ and $\mathbf{P}(z)$ are exactly the same as in Sect. 3, as they only depend on $\mathbf{R}(z)$.

4.2 Delay

Since V_C is the minimum number of work units that need to be executed by the server before the customer C can leave the system, there is a straightforward relationship between V_C and the delay D_C experienced by customer C. Specifically, the customer C will have left the system if and only if the cumulative service capacity since customer C entered the system (not counting its arrival slot) is at least V_C. So, again denoting the arrival slot of customer C by K_C, we have that

$$R_{K_C+1} + R_{K_C+2} + \ldots + R_{K_C+n} \geq V_C \quad \Leftrightarrow \quad D_C \leq n. \tag{22}$$

Therefore, for each n, the following equality holds:

$$P(R_C^{(n)} \geq V_C) = P(D_C \leq n), \tag{23}$$

where $R_C^{(n)}$ denotes $R_{K_C+1} + \ldots + R_{K_C+n}$. Multiplying by z^n and summing over all n, we then get

$$\sum_{n=0}^{\infty} P(R_C^{(n)} \geq V_C) z^n = \sum_{n=0}^{\infty} P(D_C \leq n) z^n. \tag{24}$$

The right-hand side of (24) is easily rewritten as

$$\sum_{n=0}^{\infty} P(D_C \leq n) z^n = \sum_{n=0}^{\infty} \sum_{k=0}^{n} P(D_C = k) z^n$$

$$= \sum_{k=0}^{\infty} P(D_C = k) \frac{z^k}{1-z}$$

$$= \frac{D(z)}{1-z}. \tag{25}$$

Substituting (25) into (24), we get

$$D(z) = (1-z) \sum_{n=0}^{\infty} \sum_{j=0}^{\infty} \sum_{k=j}^{\infty} P(V_C = j, R_C^{(n)} = k) z^n$$

$$= (1-z) \sum_{n=0}^{\infty} \sum_{i=1}^{m} \sum_{j=0}^{\infty} \sum_{k=j}^{\infty} P(R_C^{(n)} = k | T_{K_C+1} = i) P(V_C = j, T_{K_C+1} = i) z^n \tag{26}$$

since $R_C^{(n)}$ is independent of V_C when given T_{K_C+1}. Indeed, the cumulative service capacity $R_C^{(n)}$ depends only on the state of the Markov chain in slots $\geq K_C + 1$, while the unfinished work V_C observed by customer C is independent from those states when given the state in slot $K_C + 1$.

Given $T_{K_C+1} = i$, the random variable $R_C^{(n)}$ in (26) has a very simple conditional pgf. It can easily be checked using induction that

$$E[z^{R_C^{(n)}} I(T_{K_C+n+1} = j) | T_{K_C+1} = i] = [\mathbf{R}(z)^n]_{i,j}. \tag{27}$$

Summing over $j = 1, \ldots, m$, we thus find that the conditional pgf of $R_C^{(n)}$ given $T_{K_C+1} = i$ is simply the ith element of the column vector $\mathbf{R}(z)^n \mathbf{1}^T$.

To use this pgf in (26), we again make use of the inversion formula for probability-generating functions. This leads to

$$D(z) = \frac{1-z}{2\pi\imath} \sum_{n=0}^{\infty} \sum_{i=1}^{m} \sum_{j=0}^{\infty} \sum_{k=j}^{\infty} \oint_L \frac{[\mathbf{R}(\zeta)^n \mathbf{1}^T]_i}{\zeta^{k+1}} P(V_C = j, T_{K_C+1} = i) z^n \, \mathrm{d}\zeta,$$

where L is any counterclockwise contour around the origin for which no point on L is further from the origin than the minimum of the the radii of convergence of all elements of $\mathbf{R}(\zeta)^n \mathbf{1}^T$. We may "swap" the above infinite summations over n and k with the contour integral over L if the infinite series in the resulting integrand would be uniformly convergent. This is the case if $\forall \zeta \in L : |1/\zeta| < 1$ and the spectral radius of $z\mathbf{R}(\zeta)$ is smaller than 1. We will therefore in the sequel consider those L for which these conditions are true. In fact, since the operator norm of a matrix is never smaller than the spectral radius (by definition), we will only consider those L for which $||z\mathbf{R}(\zeta)|| < 1, \forall \zeta \in L$. Then we may write

$$D(z) = \frac{1-z}{2\pi\imath} \sum_{i=1}^{m} \sum_{j=0}^{\infty} \oint_L \frac{[(\mathbf{I} - z\mathbf{R}(\zeta))^{-1} \mathbf{1}^T]_i}{\zeta^j (\zeta - 1)} P(V_C = j, T_{K_C+1} = i) \, \mathrm{d}\zeta.$$

Similarly, if we further restrict L to be a contour where $\forall \zeta \in L : |1/\zeta| < \mathcal{R}_{[\mathbf{V}]_i}, i = 1, ..., m$, then we may swap the infinite summation over j with the contour integral over L, and we obtain

$$D(z) = \frac{1-z}{2\pi\imath} \oint_L \frac{\mathbf{V}(1/\zeta)\,(\mathbf{I} - z\mathbf{R}(\zeta))^{-1}\mathbf{1}^T}{\zeta - 1} \, \mathrm{d}\zeta. \tag{28}$$

At this point it is worth considering which poles of the integrand of (28) lie inside L. Due to our choice of L, it is easily seen that the simple pole at $\zeta = 1$ lies inside L, as do all poles of all elements of $\mathbf{V}(1/\zeta)$. It remains to be seen which singularities of $\mathbf{I} - z\mathbf{R}(\zeta)$ lie inside L. Applying Theorem 1 with $\mathbf{F}(\zeta) = \mathbf{I}$, $\mathbf{G}(\zeta) = -z\mathbf{R}(\zeta)$ and $\Gamma = L$ shows that $\mathbf{I} - z\mathbf{R}(\zeta)$ has no singularities inside L. Note in particular that $\mathbf{I} - z\mathbf{R}(0)$ is not singular.

We now substitute $\zeta = 1/\xi$ in (28) (which yields a factor $-1/\xi^2$ in the integrand) and invert the integration path L into L' but still integrate in the counterclockwise sense (which yields an extra factor -1, since the inversion of L is a clockwise path) and obtain

$$D(z) = \frac{1-z}{2\pi\imath} \oint_{L'} \frac{\mathbf{V}(\xi)\,(\mathbf{I} - z\mathbf{R}(1/\xi))^{-1}\mathbf{1}^T}{\xi(1-\xi)} \, \mathrm{d}\xi. \tag{29}$$

The poles of the integrand of (29) inside L' are now only the singularities of $\mathbf{I} - z\mathbf{R}(1/\xi)$. Note that the simple zero of the denominator at $\xi = 0$ does not cause a pole, because $\mathbf{V}(0) = \mathbf{0}$, as the unfinished work observed by a customer upon arrival can never by 0, since it at least contains the service demand of that customer. Additionally, $\mathbf{I} - z\mathbf{R}(1/\xi)$ is not singular for $\xi \to 0$ either, since that would imply that the polynomial $\det(\mathbf{I} - z\mathbf{R}(\zeta)) \to 0$ as $\zeta \to \infty$, which is impossible, as $\det(\mathbf{I} - z\mathbf{R}(\zeta))$ is a polynomial in ζ.

To evaluate (29), we use Cauchy's residue theorem and find

$$
D(z) = (1-z) \sum_{\alpha_k(z) \in \alpha(z)} \operatorname*{Res}_{\xi \to \alpha_k(z)} \left[\frac{\mathbf{V}(\xi)\,(\mathbf{I} - z\mathbf{R}(1/\xi))^{-1}\mathbf{1}^T}{\xi(1-\xi)} \right]
$$

$$
= (1-z) \sum_{\alpha_k(z) \in \alpha(z)} \frac{1}{(m_k-1)!} \lim_{\xi \to \alpha_k(z)} \frac{\mathrm{d}^{m_k-1}}{\mathrm{d}\xi^{m_k-1}} (\xi - \alpha_k(z))^{m_k}
$$

$$
\cdot \frac{\mathbf{V}(\xi)\,(\mathbf{I} - z\mathbf{R}(1/\xi))^{-1}\mathbf{1}^T}{\xi(1-\xi)}, \tag{30}
$$

where $\alpha(z)$ is the set of ξ for which $\mathbf{I} - z\mathbf{R}(1/\xi)$ is singular and m_k denotes the multiplicity of the kth singularity (i.e., the multiplicity of the kth root of the determinant). These $\alpha_k(z)$ are the multiplicative inverses of the roots for ζ of

$$
\det(\mathbf{I} - z\mathbf{R}(\zeta)) = 0, \tag{31}
$$

since $\det(\mathbf{I} - z\mathbf{R}(\zeta))$ does not approach 0 for neither $\zeta \to 0$ (as $\mathbf{I} - z\mathbf{R}(0)$ is not singular) nor $\zeta \to \infty$ (as shown earlier). Since $\det(\mathbf{I} - z\mathbf{R}(\zeta))$ is a polynomial, these $\alpha_k(z)$ and m_k can for any particular z therefore easily be calculated numerically. Unfortunately, if some of the m_k are greater than 1, then (30) is not straightforward to calculate. Luckily, there are only "very few" z for which that is the case: it can be shown that the set of z for which at least 1 of the m_k is greater than 1 forms an isolated set (see e.g. the appendix of [12] for a proof for a similar problem on a similar queueing model). For all other z, we can use l'Hôpital's rule and Jacobi's formula to simplify (30) to

$$
D(z) = \frac{z-1}{z} \sum_{\alpha_k(z) \in \alpha(z)} \frac{\alpha_k(z)}{\alpha_k(z)-1} \cdot \frac{\mathbf{V}(\alpha_k(z))\, \mathrm{adj}(\mathbf{I} - z\mathbf{R}(1/\alpha_k(z)))\mathbf{1}^T}{\operatorname{tr}\left(\mathrm{adj}(\mathbf{I} - z\mathbf{R}(1/\alpha_k(z)))\, \mathbf{R}'(1/\alpha_k(z)) \right)}. \tag{32}
$$

Substituting the expression (21) for $\mathbf{V}(z)$ in this equation does not lead to any substantial simplifications, so this is our final expression for the pgf $D(z)$ of the steady-state customer delay.

To summarize, in order to calculate $D(z)$ for any given value of z, first for each $j = 1, ..., m$, the set $[\mathcal{S}_\mathbf{R}]_j$ should be calculated. This is the set of poles for ζ of the elements of $\mathbf{R}(1/\zeta)$ on the jth column. Then $\mathcal{S}_{\tilde{\mathbf{B}}}$ should be calculated, which is the set of the zeros for ζ of $\det(\mathbf{P}(\zeta) - A(S(\zeta))\mathbf{R}(1/\zeta)\mathbf{P}(\zeta))$. Then the system of linear equations given by Eqs. (17) and (18) (with $\tilde{A}(z)$ replaced by $A(S(z))$) should be solved to determine the vector $\mathbf{C}$. Finally, the zeros for ζ of $\det(\mathbf{I} - z\mathbf{R}(\zeta))$ should be calculated, as the $\alpha_k(z)$ are the inverses of these zeros (see (31)). Then $\mathbf{V}(z)$ and $D(z)$ follow from Eqs. (21) and (32) respectively.

5   Mean Customer Delay

From the expression (32) for the pgf $D(z)$, several characteristics of the steady-state customer delay may be derived, including the mean and other moments. In this section, we briefly describe how this may be done.

It is well known that the mean of a random variable is equal to the derivative of its pgf at $z = 1$. Therefore the mean customer delay is simply equal to $D'(1)$. However, to evaluate the derivative of $D(z)$, the derivative of $\alpha_k(z)$ needs to be known. To find this, we first note that due to the fact that the $\alpha_k(z)$ are the multiplicative inverses of the roots for ζ of Eq. (31), it holds that

$$\det(\mathbf{I} - z\mathbf{R}(1/\alpha_k(z))) = 0, \tag{33}$$

for all z. Differentiating both sides of the above equation with respect to z and using Jacobi's rule, we find that

$$\mathrm{tr}\left(\mathrm{adj}(\mathbf{I} - z\mathbf{R}(1/\alpha_k(z)))\left(\mathbf{R}(1/\alpha_k(z)) - z\mathbf{R}'(1/\alpha_k(z))\frac{\alpha_k'(z)}{\alpha_k^2(z)}\right)\right) = 0. \tag{34}$$

Using (33), we can simplify this to

$$\mathrm{tr}\left(\mathrm{adj}(\mathbf{I} - z\mathbf{R}(1/\alpha_k(z)))\left(\mathbf{I} - z^2\mathbf{R}'(1/\alpha_k(z))\frac{\alpha_k'(z)}{\alpha_k^2(z)}\right)\right) = 0. \tag{35}$$

From this, we find the following expression for $\alpha_k'(z)$:

$$\alpha_k'(z) = \frac{\alpha_k^2(z)\,\mathrm{tr}\left(\mathrm{adj}(\mathbf{I} - z\mathbf{R}(1/\alpha_k(z)))\right)}{z^2\,\mathrm{tr}\left(\mathrm{adj}(\mathbf{I} - z\mathbf{R}(1/\alpha_k(z)))\,\mathbf{R}'(1/\alpha_k(z))\right)}. \tag{36}$$

Using the above equation, an expression for $D'(1)$ may be obtained by simply deriving the expression (32) for $D(z)$ and using L'Hôpital's rule. However, this is a tedious and error-prone but nevertheless straightforward calculation. To evaluate $D'(1)$ numerically for any specific choice of $A(z)$, $S(z)$ and $\mathbf{R}(z)$, an analytical expression for $D'(1)$ is in fact not needed, and a finite difference method may be used instead, where the expression (32) for $D(z)$ is used directly. That is, $D'(1)$ can be approximated very accurately by $(D(1+\epsilon) - D(1-\epsilon))/2\epsilon$ for very small ϵ. In case there are difficulties with numerical precision (which can occur if the matrix $\mathbf{R}(z)$ is large and/or contains one or more polynomials of a high degree), then either a higher-order finite difference method can be used (see e.g. [20]), or floating-point numbers with greater precision than the standard 32- or 64-bit floating point numbers can be used.

Determining the value of higher-order moments of the customer delay requires determining the value of higher-order derivatives of $D(z)$ at $z = 1$. In theory, these can be obtained in the same way as the first derivative $D'(1)$, where expressions for higher-order derivatives of $\alpha_k(z)$ may be found by repeatedly differentiating both sides of equation (36). However, the expressions involved become longer for each higher-order derivative. To determine the probability that the customer delay in an arbitrary slot in steady state is equal to a certain number, a numerical inversion method such as the one by Abate and Whitt [21] may be used.

6   Numerical Examples and Discussion

In this section, we present several illustrative numerical examples to demonstrate the behavior of the queueing model. Due to our assumption that the arrival process is uncorrelated, the most natural choice for the distribution of the number of arrivals per time slot is the Poisson distribution with parameter λ, i.e., $A(z) = e^{\lambda(z-1)}$. This arrival process is indeed what we will use in all of the numerical examples in this paper. All figures in this section were verified against Monte-Carlo simulations over 10^6 time slots.

In our first numerical example, we study the impact of the degree of correlation between the service capacities during consecutive slots on the mean customer delay in steady state. To this end, we consider a server whose service capacity can change by at most 1 from one slot to the next. The probabilities for the service capacity to increase or decrease are both equal to a certain value $\alpha \in (0, 0.5)$, with the exception that when the service capacity is 4, it cannot increase further, effectively constraining the service capacity to always be between 0 and 4 (inclusive). This gives the following expression for $\mathbf{R}(z)$:

$$\mathbf{R}(z) = \begin{bmatrix} 1-\alpha & \alpha z & 0 & 0 & 0 \\ \alpha & (1-2\alpha)z & \alpha z^2 & 0 & 0 \\ 0 & \alpha z & (1-2\alpha)z^2 & \alpha z^3 & 0 \\ 0 & 0 & \alpha z^2 & (1-2\alpha)z^3 & \alpha z^4 \\ 0 & 0 & 0 & \alpha z^3 & (1-\alpha)z^4 \end{bmatrix}. \tag{37}$$

The steady-state distribution of the service capacity is the same for all values of α and the mean service capacity is 2. However, as α decreases, the correlation between the service capacities in two consecutive slots increases.

Figure 1 shows the mean customer delay in steady state versus α for three different distributions of the service demand, namely geometric, uniform between 1 and 5 (inclusive), and deterministic, all with mean service demand 3. The arrival rate λ was taken to be 0.5, so that the traffic load is equal to 0.75. As can be seen from Fig. 1, the impact of α, and therefore of the degree of correlation between the service capacities in consecutive slots, on the mean customer delay is much greater than the specific distribution of the service demand.

For our second numerical example, we illustrate that vacations with phase-type length are simply a special case of our model, as discussed in more detail in Appendix A.

We consider a server that during the time slots where it is not in vacation has a service capacity with pgf $\hat{R}(z)$, independently from slot to slot. However, at the end of each slot, with probability 0.1 a vacation starts with shifted negative binomially distributed length with parameters $r = 5$ and $p = 0.8$. This leads to

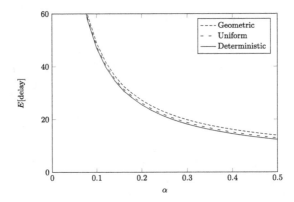

Fig. 1. Mean customer delay versus the parameter α for various distributions of the service demand.

the following expression for $\mathbf{R}(z)$:

$$\mathbf{R}(z) = \begin{bmatrix} 0.2 & 0.8 & 0 & 0 & 0 & 0 \\ 0 & 0.2 & 0.8 & 0 & 0 & 0 \\ 0 & 0 & 0.2 & 0.8 & 0 & 0 \\ 0 & 0 & 0 & 0.2 & 0.8 & 0 \\ 0 & 0 & 0 & 0 & 0.2 & 0.8 \\ 0.1\hat{R}(z) & 0 & 0 & 0 & 0 & 0.9\hat{R}(z) \end{bmatrix} \tag{38}$$

In Fig. 2, we show the mean customer delay versus the load ρ for this model for various choices of $\hat{R}(z)$, all with mean non-vacation service capacity $\hat{R}'(1) = 10$. This makes the mean service capacity μ in an arbitrarily chosen slot equal to approximately 6.15, implying that service is available approximately 61.5% of the time. Service demands are equal to 5 work units per customer. For reference, the mean customer delay for the server without any vacations and with uniformly distributed service capacities (between 0 and 20, inclusive) is also shown.

As can be seen in Fig. 2, distributions of the (non-vacation) service capacity with higher variance lead to a higher mean customer delay for all traffic intensities, as would be expected. However, for all traffic loads the server that doesn't experience vacations has a much lower mean customer delay than all the servers that do, despite the fact that the variance of the service capacity in each individual time slot is relatively high. So in this example it can again be seen that the degree of correlation between service capacities in consecutive slots has a larger impact on the mean customer demand than the variance of the service capacity distribution in each individual slot.

In our third and final numerical example, we study the impact machine defects and repairs have on the performance of a system. We consider a server that consists of two independent components, A and B. During each time slot, a working component can break with probability β, and a broken component may be repaired with probability γ. However, if both components are broken,

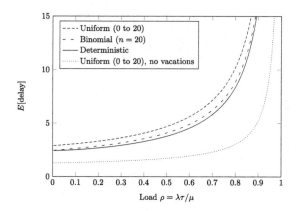

Fig. 2. Mean customer delay versus the load ρ for a server with vacations, with various distributions of the (non-vacation) service capacity. For reference, a server with uniform service capacities without vacations is also shown.

only one component may be repaired. Component A is chosen with probability α while component B is chosen with probability $1 - \alpha$. Component A adds a service capacity with pgf $\hat{R}_{A,0}(z)$ while broken and $\hat{R}_{A,1}(z)$ while working, while component B adds service capacities with pgfs $\hat{R}_{B,0}(z)$ and $\hat{R}_{B,1}(z)$ respectively. This leads to the following expression for $\mathbf{R}(z)$:

$$
\mathbf{R}(z) = \begin{bmatrix} (1-\gamma) & \gamma\alpha & \gamma(1-\alpha) & 0 \\ \beta(1-\gamma) & (1-\beta)(1-\gamma) & \beta\gamma & (1-\beta)\gamma \\ (1-\gamma)\beta & \gamma\beta & (1-\gamma)(1-\beta) & \gamma(1-\beta) \\ \beta^2 & (1-\beta)\beta & \beta(1-\beta) & (1-\beta)^2 \end{bmatrix}
$$
$$
\cdot \begin{bmatrix} \hat{R}_{A,0}(z)\hat{R}_{B,0}(z) & 0 & 0 & 0 \\ 0 & \hat{R}_{A,1}(z)\hat{R}_{B,0}(z) & 0 & 0 \\ 0 & 0 & \hat{R}_{A,0}(z)\hat{R}_{B,1}(z) & 0 \\ 0 & 0 & 0 & \hat{R}_{A,1}(z)\hat{R}_{B,1}(z) \end{bmatrix}.
$$

The following values were chosen for $\hat{R}_{A,0}(z)$, $\hat{R}_{A,1}(z)$, $\hat{R}_{B,0}(z)$, and $\hat{R}_{B,1}(z)$:

$$\hat{R}_{A,0}(z) = 0.9 + 0.1z \qquad\qquad \hat{R}_{B,0}(z) = 0.9 + 0.1z$$

$$\hat{R}_{A,1}(z) = z^3 \qquad\qquad \hat{R}_{B,1}(z) = \frac{1}{7}(1 + z + \cdots + z^6)$$

Both components add a mean service capacity of 3 when working and 0.1 when broken, but the service capacity added by component B clearly has a higher variance. Service demands are simply equal to 3 work units per customer.

The mean customer delay versus the arrival rate λ for this system is shown in Fig. 3 for various combinations of α, β and γ. A first thing to note is that the impact of α on the mean delay is relatively small, implying that the choice of which component to repair when both are broken is not very important, despite

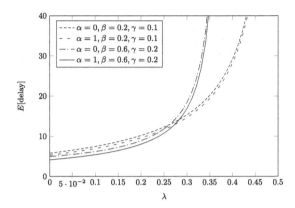

Fig. 3. Mean customer delay versus the arrival rate λ for a server with two components, each breaking with probability β and being repaired with probability γ in any slot. Only one component may be repaired per slot, component A is chosen with probability α.

the fact that while working component A always adds exactly enough service capacity to the system to serve one additional customer, while the contribution of component B is more subject to randomness.

The impact of parameters β and γ is more significant, as these parameters determine the fraction of time that machines are defect or working. The components in the server with $\beta = 0.2$ and $\gamma = 0.1$ are defect a lower fraction of the time compared to those in the server with $\beta = 0.6$ and $\gamma = 0.2$, leading to an overall mean service capacity of 1.54 and 1.14 respectively. Due to this higher mean steady-state service capacity, the server with $\beta = 0.2$ and $\gamma = 0.1$ is stable for higher traffic loads. However, for low traffic loads it can be seen that the server with $\beta = 0.6$ and $\gamma = 0.2$ has lower mean delay, despite the fact that it has a lower mean service capacity. This is because the parameters β and γ also influence how quickly the components break and get repaired. As the components break and are repaired more quickly, the correlation between the service demands in consecutive slots decreases, so does the mean service demand.

In conclusion, as in our previous numerical examples we again find that the impact of the variance of the service capacities in each individual slot, which is mostly controlled by the parameter α, is smaller than the impact of the correlation between service capacities in consecutive slots, which is mostly controlled by the parameters β and γ.

7   Conclusion

In this paper, we studied a non-classical queueing model where the customers have variable service demands and the server has a service capacity that also varies from slot to slot. Our main result was an expression for the pgf of the customer delay in steady state.

Unlike previous studies on this non-classical model, which analyzed more and more general service-capacity distributions while assuming independence between the service capacities from slot to slot, the service capacities in this paper were allowed to be correlated according to a discrete batch Markovian process. By studying several numerical examples, we found the impact of the correlation between service capacities in consecutive slots on the mean customer delay to be consistently larger than that of the variance of the service capacity in each individual slot.

Appendix

A   Vacations as a Special Case of the Model

As mentioned in our second numerical example in Sect. 6, vacations with phase-type length are simply a special case of the queuing model, and therefore do not need to be modeled separately. Indeed, consider a server for which the service capacities (without vacations) are described by the matrix generating function $\hat{\mathbf{R}}(z)$ but which can at the end of every (non-vacation) slot start a vacation with probability p with a duration that follows the discrete phase-type $PH_d(\boldsymbol{\tau}, \mathbf{T})$ distribution, i.e., the lengths of the vacations follow the same distribution as the number of steps it takes for the Markov chain with transition (block) matrix

$$\begin{bmatrix} \mathbf{T} & (\mathbf{I} - \mathbf{T})\mathbf{1} \\ \mathbf{0} & 1 \end{bmatrix}, \tag{39}$$

to go from an initial state sampled from a distribution described by the (row) vector $\boldsymbol{\tau}$ to the absorbing state. During the vacations, the available service capacity is 0.

There are several possibilities for what happens to the state of the service capacity's background Markov chain during vacations. If this background Markov chain simply keeps advancing as normal, then the system with vacations is a special case of our model where the matrix generating function of the service time is given by the block matrix

$$\mathbf{R}(z) = \begin{bmatrix} \mathbf{T} \otimes \hat{\mathbf{R}}(1) & (\mathbf{I} - \mathbf{T})\mathbf{1} \otimes \hat{\mathbf{R}}(1) \\ p\boldsymbol{\tau} \otimes \hat{\mathbf{R}}(z) & (1-p)\hat{\mathbf{R}}(z) \end{bmatrix}, \tag{40}$$

where $\otimes$ denotes the Kronecker product. Indeed, if the background Markov chain is in state $am + b$, with $0 \le b \le m - 1$, then that corresponds to the background Markov chain of the server without vacations being in state b, while a corresponds to the phase of the vacation that the server is currently in (if any).

A second possibility is if the state of the background Markov chain of the server is paused during vacations. In that case only a small modification of the expression for $\mathbf{R}(z)$ is required, and we get

$$\mathbf{R}(z) = \begin{bmatrix} \mathbf{T} \otimes \mathbf{I} & (\mathbf{I} - \mathbf{T})\mathbf{1} \otimes \mathbf{I} \\ p\boldsymbol{\tau} \otimes \hat{\mathbf{R}}(z) & (1-p)\hat{\mathbf{R}}(z) \end{bmatrix}. \tag{41}$$

A third possibility is if the state of the background Markov chain after a vacation is independent of the state before the vacation, and is instead simply equal to n with probability $[\kappa]_n$. Then $\mathbf{R}(z)$ is given by

$$\mathbf{R}(z) = \begin{bmatrix} \mathbf{T} & (\mathbf{I} - \mathbf{T})\mathbf{1} \otimes \kappa \\ p\boldsymbol{\tau} & (1-p)\hat{\mathbf{R}}(z) \end{bmatrix}. \tag{42}$$

References

1. Jin, X., Min, G., Velentzas, S., Jiang, J.: Quality-of-service analysis of queuing systems with long-range-dependent network traffic and variable service capacity. IEEE Trans. Wirel. Commun. **11**, 562–570 (2012)
2. Bosman, J.W., van der Mei, R.D., Nunez-Queija, R.: A fluid model analysis of streaming media in the presence of time-varying bandwidth. Proc. ITC **24**, 177–184 (2012)
3. Ayed, S., Sofiene, D., Nidhal, R.: Joint optimisation of maintenance and production policies considering random demand and variable production rate. Int. J. Prod. Res. **50**, 6870–6885 (2011)
4. Giri, B., Yun, W., Dohi, T.: Optimal design of unreliable production-inventory systems with variable production rate. Eur. J. Oper. Res. **162**, 372–386 (2005)
5. Boxma, O.J., Kurkova, I.A.: The M/G/1 queue with two service speeds. Adv. Appl. Probab. **33**, 520–540 (2001)
6. Halfin, S.: Steady-state distribution for the buffer content of an M/G/1 queue with varying service rate. SIAM J. Appl. Math. **23**, 356–363 (1972)
7. Mahabhashyam, S.R., Gautam, N.: On queues with Markov modulated service rates. Queueing Syst. **51**, 89–113 (2005)
8. Bruneel, H., Wittevrongel, S., Claeys, D., Walraevens, J.: Discrete-time queues with variable service capacity: a basic model and its analysis. Ann. Oper. Res. **239**(2), 359–380 (2016)
9. Walraevens, J., Bruneel, H., Claeys, D., Wittevrongel, S.: The discrete-time queue with geometrically distributed service capacities revisited. In: Dudin, A., De Turck, K. (eds.) ASMTA 2013. LNCS, vol. 7984, pp. 443–456. Springer, Heidelberg (2013)
10. Bruneel, H., Rogiest, W., Walraevens, J., Wittevrongel, S.: On queues with general service demands and constant service capacity. In: Norman, G., Sanders, W. (eds.) QEST 2014. LNCS, vol. 8657, pp. 210–225. Springer, Cham (2014)
11. Yao, Y., Wei-Chung Miao, D.: Sample-path analysis of general arrival queueing systems with constant amount of work for all customers. Queueing Syst. **76**(3), 283–308 (2014)
12. De Muynck, M., Wittevrongel, S., Bruneel, H.: Analysis of discrete-time queues with general service demands and finite-support service capacities. Ann. Oper. Res. **252**(1), 3–28 (2017)
13. De Muynck, M., Bruneel, H., Wittevrongel, S.: Delay analysis of a queue with general service demands and phase-type service capacities. In: van Do, T., Takahashi, Y., Yue, W., Nguyen, V.-H. (eds.) Queueing Theory and Network Applications. AISC, vol. 383, pp. 29–39. Springer, Cham (2016). https://doi.org/10.1007/978-3-319-22267-7_3
14. De Muynck, M., Bruneel, H., Wittevrongel, S.: Analysis of a discrete-time queue with general service demands and phase-type service capacities. J. Ind. Manag. Optim. **13**(4), 1901–1926 (2017)

15. Claeys, D., Walraevens, J., Laevens, K., Steyaert, B., Bruneel, H.: A batch-service queueing model with a discrete batch Markovian arrival process. In: Al-Begain, K., Fiems, D., Knottenbelt, W.J. (eds.) ASMTA 2010. LNCS, vol. 6148, pp. 1–13. Springer, Heidelberg (2010)
16. Wang, Y.C., Cai, D.L., Chiang, L.H., Hu, C.W.: Elucidating the short term loss behavior of Markovian-modulated batch-service queueing model with discrete-time batch Markovian arrival process. Math. Probl. Eng. **2014**, 1–10 (2014). Article ID 814810
17. Ellis, R.L., Gohberg, I.: Orthogonal Systems and Convolution Operators. Operator Theory: Advances and Applications, vol. 140. Birkhäuser Verlag, Basel (2003). https://doi.org/10.1007/978-3-0348-8045-9
18. Chaudhry, M.L., Singh, G., Gupta, U.C.: A simple and complete computational analysis of MAP/R/1 queue using roots. Methodol. Comput. Appl. Probab. **15**(3), 563–582 (2013)
19. Bruneel, H., Kim, B.G.: Discrete-Time Models for Communication Systems Including ATM. Kluwer Academic, Boston (1993)
20. Fornberg, B.: Generation of finite difference formulas on arbitrarily spaced grids. Math. Comput. **51**, 699–706 (1988)
21. Abate, J., Whitt, W.: Numerical inversion of probability generating functions. Oper. Res. Lett. **12**(4), 245–251 (1992)

Fluid Polling System with Markov Modulated Load and Gated Discipline

Zsolt Saffer[1], Miklós Telek[2], and Gábor Horváth[3(✉)]

[1] Institute of Statistics and Mathematical Methods in Economics,
Vienna University of Technology, Vienna, Austria
`zsolt.saffer@tuwien.ac.at`
[2] MTA-BME Information Systems Research Group, Budapest University
of Technology and Economics, Magyar Tudósok Körútja 2, Budapest 1117, Hungary
[3] Department of Networked Systems and Services, Budapest University
of Technology and Economics, Magyar Tudósok Körútja 2, Budapest 1117, Hungary
`{telek,ghorvath}@hit.bme.hu`

Abstract. In this paper we provide an analysis for fluid polling models
with Markov modulated load and gated discipline. The fluid arrival to
the stations is modulated by a common continuous-time Markov chain.
The fluid is removed at the stations during the service period by a station
dependent constant rate.

We build partly on the methods used previously in the analysis of
fluid vacation models with gated discipline. We establish steady-state
relationships on Laplace transform level regarding the joint distribution
of the fluid levels at the stations and the state of the modulating Markov
chain among different characteristic epochs including start and end of
the service at each station. We derive the steady-state vector Laplace
transform of the fluid levels at the stations at arbitrary epoch and its
mean.

Keywords: Queueing theory · Fluid model · Polling system
Gated discipline

1 Introduction

In fluid queueing models the work arrives on a continuous manner, i.e., fluid
flows into the buffer instead of customer arrivals. Such models can be used as
the limit for the workload in the analysis of regular queueing systems, for example
in Heavy-Traffic (HT) analysis or stability analysis [1,2].

The Markov modulated fluid queues have been analyzed by several authors
using matrix analytic methods, see, e.g., [3,4].

The first paper relevant to fluid polling model is the paper from Czerniak and
Yechiali [5]. They analyzed a fluid polling model with constant load and service
rate. The only non-deterministic part of their model is the switchover time.

Fluid vacation models with Markov modulated load have been analyzed in
the subsequent papers [6–8]. The authors studied the fluid vacation models with

© Springer International Publishing AG, part of Springer Nature 2018
Y. Takahashi et al. (Eds.): QTNA 2018, LNCS 10932, pp. 86–102, 2018.
https://doi.org/10.1007/978-3-319-93736-6_6

gated discipline and with exhaustive discipline under negative fluid rate during service. The analysis of the exhaustive fluid vacation model has been extended to the case of the non-negative fluid rate during service in [9].

This work is a natural continuation of the above research line on fluid vacation models in which we extend the analysis of fluid gated vacation model to the corresponding fluid polling system. The contribution of this work is the extension of the analysis of fluid gated vacation model with Markov modulated load to the fluid polling system. However, we build only partly on the methods used in the analysis of fluid vacation model with gated discipline. We establish steady-state relationships on Laplace transform (LT) level regarding the joint distribution of the fluid levels at the stations and the state of the modulating Markov chain among different characteristic epochs, like start and end of the service at each station. We derive the steady-state vector LT of the fluid levels at the stations at arbitrary epoch and its mean.

The rest of the paper is organized as follows. Section 2 gives the description and the stability criterion of the model. The analysis of the steady-state fluid levels at characteristic epochs follows in Sect. 3. Section 4 provides the analysis of the steady-state fluid levels at arbitrary epoch and its mean.

2   Model and Notation

2.1   Model Description

We consider a fluid polling model with Markov modulated load and gated discipline. The polling system consists of N stations. Each station has an infinite fluid buffer.

A common continuous-time Markov chain (CTMC) ($\Omega(t)$ for $t \geq 0$) with state space $\Omega = \{1, \ldots, L\}$ modulates the arriving fluid flows at the station. The generator of this background CTMC is denoted by $\mathbf{Q}$. The input fluid rates at station i are specified by diagonal fluid input rate matrix $\mathbf{R}_i$, for $i \in \{1, \ldots, N\}$. If the background CTMC is in state j ($\Omega(t) = j$) then fluid flows into the buffer of station i at rate $r_i(j)$ for $j \in \{1, \ldots, L\}$ and $i \in \{1, \ldots, N\}$. When the server visits station i it removes fluid from its fluid buffer at finite rate $d_i > 0$ for $i \in \{1, \ldots, N\}$. Consequently, when the server visits station i and the overall Markov chain is in state j ($\Omega(t) = j$) then the fluid level of the buffer of station i changes at rate $r_i(j) - d_i$ otherwise it changes at rate $r_i(j)$ due to the lack of service. The length of the server's visit at station i in the polling model is determined by the service discipline applied at that station. In this work we consider the gated discipline. Under gated discipline only the fluid is removed during the server visit at station i, which is present at the station already upon the server arrival. The cycle time (or simple cycle) is the time between two consecutive visits of the server to the same station. In this paper, if not stated otherwise then we understand the station index i as $mod(N)$, i.e. whenever it reaches N it continues by 1. The switchover time from station i to the next station in the consecutive cycles is independent and identically distributed (i.i.d.). The probability distribution function (pdf) of the switchover time from station i, the corresponding Laplace transform (LT) and mean is denoted by $\sigma_i(t)$ and $\sigma_i^*(s)$, σ_i, respectively. We consider non-zero switchover-times model, and we use the notation $\sigma = \sum_{i=1}^N \sigma_i$. We set the following assumptions on the fluid polling model:

- **A.1** The generator matrix $\mathbf{Q}$ of the modulating CTMC is irreducible.
- **A.2** The fluid rates $r_i(j)$ are positive and finite, i.e. $r_i(j > 0$ for $j \in \{1, \ldots, L\}$ and $i \in \{1, \ldots, N\}$.

Remark 1. The case of independent fluid inputs is also included by the approach with one common modulating CTMC as special case. In that case $\mathbf{Q} = \oplus_{i=1}^N \hat{\mathbf{Q}}_i$ and $\mathbf{R}_i = (\otimes_{k=1}^{i-1} \mathbf{I}) \otimes \hat{\mathbf{R}}_i \otimes (\otimes_{k=i+1}^N \mathbf{I})$, where $\hat{\mathbf{Q}}_i$ and $\hat{\mathbf{R}}_i$ denote the independent generator and the fluid input rate matrix of station i, for $i \in \{1, \ldots, N\}$, and $\otimes$ and $\oplus$ denote the Kronecker product and Kronecker sum operations, respectively.

Let π be the stationary probability vector of the modulating Markov chain. Due to assumption **A.1**, $\pi\mathbf{Q} = 0$ and $\pi\mathbf{e} = 1$ uniquely determine π, where $\mathbf{e}$ is the $L \times 1$ unit column vector. The stationary fluid flow rate and the utilization at station i, λ_i and ρ_i, respectively, can be given for $i \in \{1, \ldots, N\}$ as

$$\lambda_i = \pi\mathbf{R}_i\mathbf{e} \text{ and } \rho_i = \frac{\lambda_i}{d_i}, \tag{1}$$

and the total utilization is

$$\rho = \sum_{i=1}^N \rho_i. \tag{2}$$

The arrival instant of the server to station i is called i-polling epoch. Similarly, the time instant when the server departs from station i is called i-departure epoch.

For the j, l element of the matrix $\mathbf{Z}$ the notation $\mathbf{Z}_{j,l}$ is used. Furthermore, $[\mathbf{z}_i]_j$ denote the j-th element of vector $\mathbf{z}_i$. When there is a set of random variables characterized by one (two) parameters, e.g., Y_n ($Y_{k,n}$), then the n (k, n) element of its vector (matrix) LT is $E(e^{-sY_n})$ ($E(e^{-sY_{k,n}})$). When $\mathbf{X}^*(v)$, $Re(v) \geq 0$ is a matrix LT, $\mathbf{X}^{(k)}$ denotes its k-th ($k \geq 1$) moment, i.e., $\mathbf{X}^{(k)} = (-1)^k \frac{d^k}{ds^k}\mathbf{X}^*(v)|_{v=0}$ and $\mathbf{X}$ denotes its value at $s = 0$, i.e., $\mathbf{X} = \mathbf{X}^*(0)$. Similarly when $\mathbf{x}^*(v)$, $Re(v) \leq 0$ is a vector LT, $\mathbf{x}^{(k)}$ denotes its k-th ($k \geq 1$) moment, i.e., $\mathbf{x}^{(k)} = (-1)^k \frac{d^k}{ds^k}\mathbf{x}^*(v)|_{v=0}$ and $\mathbf{x}$ denotes its value at $s = 0$, i.e., $\mathbf{x} = \mathbf{x}^*(0)$.

2.2 Stability

We apply a workload argument to get a necessary condition of the stability. The amount of work flowing to station i during a time unit is equal to its utilization, ρ_i. The necessary condition of the stability is that the total amount of work flowing to all stations during a time unit must be less than the work-amount of that time unit, which is 1. Therefore the necessary condition of the stability is given as

$$\rho < 1. \tag{3}$$

Remark 2. If the system would limit the work which could be done on average, i.e., when less then 1 work-amount could be done during a time unit, then further restrictions were needed for the sufficiency. However, the gated discipline is "unlimited", since it does not set any load-independent limit on the work-amount, which could be performed during a service period. Therefore the above necessary condition is also a sufficient one for the stability of the system.

3 The Steady-State Fluid Levels at Polling Epochs

3.1 Transient Analysis of the Accumulated Fluid

In this section, we consider the joint distribution of the accumulated amount of fluid entering into the individual stations during time $t \geq 0$. We derive the joint LT of the accumulated fluid levels flowed into the stations and the state of the common modulated Markov chain as a function of time.

Let $Y_i(t) \in \mathbb{R}^+$ denote the accumulated amount of fluid entering into station i until time t for $i \in \{1, \ldots, N\}$. Using the notation $\bar{y} = (y_1, \ldots, y_N)$ let the transition density matrix $\mathbf{A}(t, \bar{y})$ be composed by its elements $\mathbf{A}_{j,k}(t, \bar{y})$ as

$$\mathbf{A}_{j,k}(t, \bar{y}) = \frac{\partial}{\partial y_1} \ldots \frac{\partial}{\partial y_N}$$
$$Pr(\Omega(t) = k, Y_1(t) < y_1, \ldots Y_N(t) < y_N | \Omega(0) = j, Y_1(0) = \ldots = Y_N(0) = 0).$$

The fluid level is zero at each station i at $t = 0$ ($Y_i(0) = 0$) with probability 1. Hence the transition density matrix for $t = 0$ is given as

$$\mathbf{A}(0, y_1, \ldots, y_N) = \delta(y_1) \ldots \delta(y_N)\mathbf{I}, \tag{4}$$

where $\delta(y)$ denotes the unit impulse function at $y = 0$, whose LT is 1. Furthermore the accumulated amount of fluids are greater than zero for $t > 0$ at every stations ($Y_i(t) > 0$, for $i \in \{1, \ldots, N\}$) due to assumption **A.2**. It follows that

$$\mathbf{A}(t, y_1, \ldots, y_{i-1}, 0, y_{i+1}, \ldots, y_N) = \mathbf{0}, \quad t > 0, i \in \{1, \ldots, N\}, \tag{5}$$

where $\mathbf{0}$ denotes the $L \times L$ zero matrix. We also use the notation $\bar{v} = (v_1, \ldots, v_N)$ and we define several LTs of matrix $\mathbf{A}(t, \bar{y})$ as

$$\mathbf{A}^*(s, \bar{y}) = \int_{t=0}^{\infty} \mathbf{A}(t, y_1, \ldots, y_N)e^{-st}dt,$$

$$\mathbf{A}^{N*}(t, \bar{v}) = \int_{y_1=0}^{\infty} \ldots \int_{y_N=0}^{\infty} \mathbf{A}(t, y_1, \ldots, y_N)e^{-v_1 y_1} \ldots e^{-v_N y_N} \, dy_N \ldots dy_1,$$

$$\mathbf{A}^{(N+1)*}(s, \bar{v}) = \int_{y_1=0}^{\infty} \ldots \int_{y_N=0}^{\infty} \mathbf{A}^*(s, y_1, \ldots, y_N)e^{-v_1 y_1} \ldots e^{-v_N y_N} \, dy_N \ldots dy_1,$$

and

$$\mathbf{A}^{(N)*}(s, v_1, \ldots, v_{i-1}, 0, v_{i+1}, \ldots, v_N) =$$
$$\int_{y_1=0}^{\infty} \cdots \int_{y_{i-1}=0}^{\infty} \int_{y_{i+1}=0}^{\infty} \cdots \int_{y_N=0}^{\infty} \mathbf{A}^*(s, y_1, \ldots, y_{i-1}, 0, y_{i+1} \ldots y_N)$$
$$e^{-v_1 y_1} \ldots e^{-v_{i-1} y_{i-1}} e^{-v_{i+1} y_{i+1}} \ldots e^{-v_N y_N} \, dy_N \ldots dy_{i+1} dy_{i-1} \ldots dy_1,$$

where the coefficients of $*$ in the superscript of matrix $\mathbf{A}$ denotes the number of LTs.

Proposition 1. *In the fluid polling model the joint matrix LT of the accumulated amount of fluid entering in interval $(0, t]$ can be expressed as*

$$\mathbf{A}^{(N)*}(t, \bar{v}) = e^{-t\left(\sum_{i=1}^{N} \mathbf{R}_i v_i - \mathbf{Q}\right)}. \tag{6}$$

Proof. The Markov process $\{\Omega(t), Y_1(t), \ldots, Y_N(t)\}$ describes a homogenous first order fluid model. Its transient behavior can be characterized by forward Kolmogorov equations as

$$\frac{\partial}{\partial t} \mathbf{A}(t, \bar{y}) + \frac{\partial}{\partial y_1} \mathbf{A}(t, \bar{y}) \mathbf{R}_1 + \ldots + \frac{\partial}{\partial y_N} \mathbf{A}(t, \bar{y}) \mathbf{R}_N = \mathbf{A}(t, \bar{y}) \mathbf{Q}. \tag{7}$$

and with initial conditions (4) and (5). Taking the LT of (7) with respect to t yields

$$\mathbf{A}^*(s, \bar{y}) s - \mathbf{A}(0, \bar{y}) + \frac{\partial}{\partial y_1} \mathbf{A}^*(s, \bar{y}) \mathbf{R}_1 + \ldots + \frac{\partial}{\partial y_N} \mathbf{A}^*(s, \bar{y}) \mathbf{R}_N = \mathbf{A}^*(s, \bar{y}) \mathbf{Q}. \tag{8}$$

Now taking the LT of (8) with respect to $y_1, \ldots, y_N$ we have

$$\mathbf{A}^{(N+1)*}(s, \bar{v}) s - \mathbf{A}^{(N)*}(0, \bar{v})$$
$$+ \left(\mathbf{A}^{(N+1)*}(s, \bar{v}) v_1 - \mathbf{A}^{(N)*}(s, 0, v_2, \ldots, v_N) \right) \mathbf{R}_1 + \ldots$$
$$+ \left(\mathbf{A}^{(N+1)*}(s, \bar{v}) v_N - \mathbf{A}^{(N)*}(s, v_1, \ldots, v_{N-1}, 0) \right) \mathbf{R}_N$$
$$= \mathbf{A}^{(N+1)*}(s, \bar{v}) \mathbf{Q}. \tag{9}$$

Applying (4) and (5) in (9) gives

$$\mathbf{A}^{(N+1)*}(s, \bar{v}) s - \mathbf{I} + \mathbf{A}^{(N+1)*}(s, \bar{v}) \mathbf{R}_1 v_1 + \ldots + \mathbf{A}^{(N+1)*}(s, \bar{v}) \mathbf{R}_N v_N$$
$$= \mathbf{A}^{(N+1)*}(s, \bar{v}) \mathbf{Q}. \tag{10}$$

After rearranging (10) we get

$$\mathbf{A}^{(N+1)*}(s, \bar{v}) = (\mathbf{I} s + \mathbf{R}_1 v_1 + \ldots + \mathbf{R}_N v_N - \mathbf{Q})^{-1}. \tag{11}$$

Taking the inverse Laplace transform of (11) with respect to s results in the statement of the proposition.

3.2   The Governing Equations of the System at Polling and Departure Epochs

Let $X_i(t) \in \mathbb{R}^+$ denote the actual level of the fluid buffer at station i at time t for $i \in \{1, \ldots, N\}$. Let $t_i^f(\ell)$ be the time of the i-polling epoch in the ℓ-th cycle for $\ell \geq 1$ and $i = \{1, \ldots, N\}$. We use the notation $\bar{x} = (x_1, \ldots, x_N)$. We define the joint densities of the fluid levels at the stations and the state of the modulating Markov chain at the i-polling epoch in the ℓ-th cycle, for $\ell \geq 1$ and $i = \{1, \ldots, N\}$, the $1 \times L$ vector $\mathbf{f}_i(\ell, \bar{x})$ by its elements as

$$[\mathbf{f}_i(\ell, \bar{x})]_j = \frac{\partial}{\partial x_1} \cdots \frac{\partial}{\partial x_N}$$
$$Pr(\Omega(t_i^f(\ell)) = j, X_1(t_i^f(\ell)) < x_1, \ldots X_N(t_i^f(\ell)) < x_N), \; j \in \Omega.$$

The steady-state counterpart of the vector $\mathbf{f}_i(\ell, \bar{x})$ is defined as

$$\mathbf{f}_i(\bar{x}) = \lim_{\ell \to \infty} \mathbf{f}_i(\ell, \bar{x}),$$

and its LT is given as

$$\mathbf{f}_i^{(N)*}(\bar{v}) = \int_{x_1=0}^{\infty} \cdots \int_{x_N=0}^{\infty} \mathbf{f}_i(\bar{x}) e^{-v_1 x_1} \ldots e^{-v_N x_N} dx_N \ldots dx_1.$$

Analogously let $t_i^m(\ell)$ be the time of the i-departure epoch in the ℓ-th cycle for $\ell \geq 1$ and $i = \{1, \ldots, N\}$. We define the joint densities of the fluid levels at the stations and the state of the modulating Markov chain at the i-departure epoch in the ℓ-th cycle, for $\ell \geq 1$ and $i = \{1, \ldots, N\}$, the $1 \times L$ vector $\mathbf{m}_i(\ell, \bar{x})$ by its elements as

$$[\mathbf{m}_i(\ell, \bar{x})]_j = \frac{\partial}{\partial x_1} \cdots \frac{\partial}{\partial x_N}$$
$$Pr(\Omega(t_i^m(\ell)) = j, X_1(t_i^m(\ell)) < x_1, \ldots X_N(t_i^m(\ell)) < x_N), \; j \in \Omega.$$

The steady-state joint densities of the fluid levels at the stations and the state of the modulating Markov chain at the i-departure epoch are defined as

$$\mathbf{m}_i(\bar{x}) = \lim_{\ell \to \infty} \mathbf{m}_i(\ell, \bar{x}),$$

and its LT is given as

$$\mathbf{m}_i^{(N)*}(\bar{v}) = \int_{x_1=0}^{\infty} \cdots \int_{x_N=0}^{\infty} \mathbf{m}_i(\bar{x}) e^{-v_1 x_1} \ldots e^{-v_N x_N} dx_N \ldots dx_1.$$

We define a notation for substituting the multivariate $L \times L$ matrix function $\mathbf{H}(\bar{v})$ into the defining integral of the LT $\mathbf{f}_i^{(N)*}(\bar{v})$ as

$$\mathbf{f}_i^{(N)*}(v_1, \ldots, v_{i-1}, \mathbf{H}(\bar{v}), v_{i+1}, \ldots, v_N) = \tag{12}$$
$$\int_{x_1=0}^{\infty} \cdots \int_{x_N=0}^{\infty} \mathbf{f}_i(\bar{x}) e^{-v_1 x_1} \ldots e^{-v_{i-1} x_{i-1}} e^{-\mathbf{H}(\bar{v}) x_i} e^{-v_{i+1} x_{i+1}} \ldots e^{-v_N x_N} dx_N \ldots dx_1.$$

Theorem 1. *The governing equations of the stable fluid polling model with gated discipline in terms of the steady-state joint vector LTs of the fluid levels at the stations at the i-polling and i-departure epochs for $i \in \{1, \ldots, N\}$ are given as*

– *for the transition* $\mathbf{f}_i \to \mathbf{m}_i$

$$\mathbf{m}_i^{(N)*}(\bar{v}) = \mathbf{f}_i^{(N)*}(v_1, \ldots, v_{i-1}, \frac{\sum_{i=1}^{N} \mathbf{R}_i v_i - \mathbf{Q}}{d_i}, v_{i+1}, \ldots, v_N), \qquad (13)$$

– *and for the transition* $\mathbf{m}_i \to \mathbf{f}_{i+1}$

$$\mathbf{f}_{i+1}^{(N)*}(\bar{v}) = \mathbf{m}_i^{(N)*}(\bar{v}))\sigma_i^*(\sum_{i=1}^{N} \mathbf{R}_i v_i - \mathbf{Q}). \qquad (14)$$

Proof. Due to the gated service discipline the fluid level at station i at i-departure epoch equals the level of the fluid arriving during the service duration of station i. The fluid level at stations $j \neq i$ at i-departure epoch is the sum of the fluid level at the previous i-polling epoch and the fluid arrived in between. If the fluid level at station i at i-polling epoch equals $\xi_i > 0$ then service duration is $\frac{\xi_i}{d_i}$ due to the gated discipline. Accordingly we can express $[\mathbf{m}_i(\bar{x})]_k$ as

$$[\mathbf{m}_i(\bar{x})]_k = \sum_{j=1}^{L} \int_{\xi_i=0}^{\infty} \int_{y_1=0}^{x_1} \cdots \int_{y_{i-1}=0}^{x_{i-1}} \int_{y_{i+1}=0}^{x_{i+1}} \cdots \int_{y_N=0}^{x_N}$$
$$[\mathbf{f}_i(x_1 - y_1, \ldots, x_{i-1} - y_{i-1}, \xi_i, x_{i+1} - y_{i+1}, \ldots, x_N - y_N)]_j$$
$$\mathbf{A}_{jk}(\frac{\xi_i}{d_i}, y_1, \ldots, y_{i-1}, x_i, y_{i+1}, \ldots, y_N) dy_N \ldots dy_{i+1} dy_{i-1} \ldots dy_1 d\xi_i.$$

Changing to vector and matrix notation results in

$$\mathbf{m}_i(\bar{x}) = \int_{\xi_i=0}^{\infty} \int_{y_1=0}^{x_1} \cdots \int_{y_{i-1}=0}^{x_{i-1}} \int_{y_{i+1}=0}^{x_{i+1}} \cdots \int_{y_N=0}^{x_N}$$
$$\mathbf{f}_i(x_1 - y_1, \ldots, x_{i-1} - y_{i-1}, \xi_i, x_{i+1} - y_{i+1}, \ldots, x_N - y_N)$$
$$\mathbf{A}(\frac{\xi_i}{d_i}, y_1, \ldots, y_{i-1}, x_i, y_{i+1}, \ldots, y_N) dy_N \ldots dy_{i+1} dy_{i-1} \ldots dy_1 d\xi_i.$$

Using the convolution property of the LT, the LT of $\mathbf{m}_i(\bar{x})$ with respect to $\bar{x}$ can be given as

$$\mathbf{m}_i^{(N)*}(\bar{v}) = \int_{\xi_i=0}^{\infty} \mathbf{f}_i^{(N-1)*}(v_1, \ldots, v_{i-1}, \xi_i, v_{i+1}, \ldots, v_N) \mathbf{A}^{(N)*}(\frac{\xi_i}{d_i}, \bar{v}) d\xi_i. \quad (15)$$

Applying (6) in (15) yields

$$\mathbf{m}_i^{(N)*}(\bar{v}) = \int_{\xi_i=0}^{\infty} \mathbf{f}_i^{(N-1)*}(v_1, \ldots, v_{i-1}, \xi_i, v_{i+1}, \ldots, v_N) e^{-\frac{\xi_i}{d_i}(\sum_{i=1}^{N} \mathbf{R}_i v_i - \mathbf{Q})} d\xi_i. \quad (16)$$

The first statement of the theorem comes by observing that the right hand side of (16) is an LT with respect to ξ_i and applying the notation (12).

The fluid level at any station j at $i + 1$-polling epoch is the sum of the fluid level at the previous i-departure epoch and the fluid arrived in between. Therefore we have

$$[\mathbf{f}_{i+1}(\bar{x})]_k = \sum_{j=1}^{L} \int_{t=0}^{\infty} \int_{y_1=0}^{x_1} \cdots \int_{y_N=0}^{x_N} [\mathbf{m}_i(x_1 - y_1, \ldots, x_N - y_N)]_j$$
$$\mathbf{A}_{jk}(t, y_1, \ldots, y_N)\sigma_i(t)dy_N \ldots dy_1 dt. \qquad (17)$$

Changing (17) to matrix notation and using the convolution property of LT we get

$$\mathbf{f}_{i+1}^{(N)*}(\bar{v}) = \int_{t=0}^{\infty} \mathbf{m}_i^{(N)*}(\bar{v}) \mathbf{A}^{(N)*}(t, \bar{v})\sigma_i(t)dt. \qquad (18)$$

Applying (6) in (18) and rearrangement leads to

$$\mathbf{f}_{i+1}^{(N)*}(\bar{v}) = \mathbf{m}_i^{(N)*}(\bar{v})) \int_{t=0}^{\infty} e^{-t(\sum_{i=1}^{N} \mathbf{R}_i v_i - \mathbf{Q})}\sigma_i(t)dt. \qquad (19)$$

The second statement of the theorem comes by observing that on the r.h.s. of (19) there is an LT with respect to t. □

3.3 The Steady-State Vector Moments of the Fluid Levels at Polling Epochs

Corollary 1. *The relation for the transition* $\mathbf{f}_i \rightarrow \mathbf{f}_{i+1}$, *for* $i \in \{1, \ldots, N\}$ *in the stable fluid polling model with gated discipline are given as*

$$\mathbf{f}_{i+1}^{(N)*}(\bar{v}) = \mathbf{f}_i^{(N)*}\left(v_1, \ldots, v_{i-1}, \frac{\sum_{i=1}^{N} \mathbf{R}_i v_i - \mathbf{Q}}{d_i}, v_{i+1}, \ldots, v_N\right) \sigma_i^*\left(\sum_{i=1}^{N} \mathbf{R}_i v_i - \mathbf{Q}\right),$$
$$(20)$$

Proof. The corollary comes by applying (13) in (14). □

We define the joint moments of the fluid levels at the stations as

$$\mathbf{f}_i^{(j_1, \ldots, j_N)} = (-1)^{\sum_{m=1}^{N} j_m} \frac{\partial^{j_1}}{\partial v_1^{j_1}} \cdots \frac{\partial^{j_N}}{\partial v_N^{j_N}} \mathbf{f}_i^{(N)*}(v_1, \ldots, v_N)\bigg|_{v_1 = \cdots = v_N = 0}.$$

Furthermore, we define the following quantities

$$\mathbf{H}_k^{(j_1, \ldots, j_N)} = (-1)^{\sum_{m=1}^{N} j_m} \frac{1}{k!} \frac{\partial^{j_1}}{\partial v_1^{j_1}} \cdots \frac{\partial^{j_N}}{\partial v_N^{j_N}} \left(\frac{\mathbf{Q} - \sum_{i=1}^{N} \mathbf{R}_i v_i}{d_i}\right)^k \bigg|_{v_1 = \cdots = v_N = 0}$$

$$\sigma_i^{(j_1, \ldots, j_N)} = (-1)^{\sum_{m=1}^{N} j_m} \frac{\partial^{j_1}}{\partial v_1^{j_1}} \cdots \frac{\partial^{j_N}}{\partial v_N^{j_N}} \sigma_i^*\left(\sum_{i=1}^{N} \mathbf{R}_i v_i - \mathbf{Q}\right)\bigg|_{v_1 = \cdots = v_N = 0}$$

Corollary 2. *The joint moments of the fluid levels at the stations can be determined from the following approximate system of linear equations*

$$
\mathbf{f}_{i+1}^{(j_1,\dots,j_N)} = \sum_{j_{1,1}+\dots+j_{1,3}=j_1} \binom{j_1}{j_{1,1},j_{1,2},j_{1,3}} \cdots \sum_{j_{N,1}+\dots+j_{N,3}=j_N} \binom{j_N}{j_{N,1},j_{N,2},j_{N,3}}
$$

$$
\sum_{k=0}^{K-j_{i,1}} \mathbf{f}_i^{(j_{1,1},\dots,j_{i-1,1},j_{i,1}+k,j_{i+1,1},\dots,j_{N,1})}\mathbf{H}_k^{(j_{1,2},\dots,j_{N,2})}\sigma_i^{(j_{1,3},\dots,j_{N,3})}, \quad (21)
$$

where $j_1,\dots,j_N = 0,\dots K$ *and* $i \in \{1,\dots,N\}$.

Proof. Taking $(-1)^{\sum_{m=1}^{N} j_m} \dfrac{\partial^{j_1}}{\partial v_1^{j_1}}\cdots\dfrac{\partial^{j_N}}{\partial v_N^{j_N}}$ on (20) and setting $v_1=\dots=v_N=0$ gives

$$
\mathbf{f}_{i+1}^{(j_1,\dots,j_N)} = (-1)^{\sum_{m=1}^{N} j_m}\frac{\partial^{j_1}}{\partial v_1^{j_1}}\cdots\frac{\partial^{j_N}}{\partial v_N^{j_N}}\int_{y_i=0}^{\infty}\mathbf{f}_i^{(N-1)*}(v_1,\dots,v_{i-1},y_i,v_{i+1},\dots,v_N)
$$

$$
e^{-y_i\frac{\sum_{i=1}^{N}\mathbf{R}_i v_i - \mathbf{Q}}{d_i}}dy_i\ \sigma_i^*(\sum_{i=1}^{N}\mathbf{R}_i v_i - \mathbf{Q})\Bigg|_{v_1=\dots=v_N=0}. \quad (22)
$$

Rearranging (22) leads to

$$
\mathbf{f}_{i+1}^{(j_1,\dots,j_N)} = (-1)^{\sum_{m=1}^{N} j_m}\frac{\partial^{j_1}}{\partial v_1^{j_1}}\cdots\frac{\partial^{j_N}}{\partial v_N^{j_N}}\int_{y_i=0}^{\infty}\mathbf{f}_i^{(N-1)*}(v_1,\dots,v_{i-1},y_i,v_{i+1},\dots,v_N)
$$

$$
\sum_{k=0}^{\infty}\frac{y_i^k}{k!}\left(\frac{\mathbf{Q}-\sum_{i=1}^{N}\mathbf{R}_i v_i}{d_i}\right)^k dy_i\ \sigma_i^*(\sum_{i=1}^{N}\mathbf{R}_i v_i - \mathbf{Q})\Bigg|_{v_1=\dots=v_N=0}
$$

$$
= (-1)^{\sum_{m=1}^{N} j_m}\frac{\partial^{j_1}}{\partial v_1^{j_1}}\cdots\frac{\partial^{j_N}}{\partial v_N^{j_N}}\sum_{k=0}^{\infty}(-1)^k\frac{\partial^k}{\partial v_i^k}\mathbf{f}_i^{(N)*}(v_1,\dots,v_N)
$$

$$
\frac{1}{k!}\left(\frac{\mathbf{Q}-\sum_{i=1}^{N}\mathbf{R}_i v_i}{d_i}\right)^k \sigma_i^*(\sum_{i=1}^{N}\mathbf{R}_i v_i - \mathbf{Q})\Bigg|_{v_1=\dots=v_N=0}
$$

$$
= \sum_{j_{1,1}+\dots+j_{1,3}=j_1} \binom{j_1}{j_{1,1},j_{1,2},j_{1,3}} \cdots \sum_{j_{N,1}+\dots+j_{N,3}=j_N} \binom{j_N}{j_{N,1},j_{N,2},j_{N,3}}
$$

$$
\sum_{k=0}^{\infty} \mathbf{f}_i^{(j_{1,1},\dots,j_{i-1,1},j_{i,1}+k,j_{i+1,1},\dots,j_{N,1})}\mathbf{H}_k^{(j_{1,2},\dots,j_{N,2})}\sigma_i^{(j_{1,3},\dots,j_{N,3})}. \quad (23)
$$

The statement of the corollary comes by applying a truncation at K in the order of the moments. □

The truncation applied in corollary 2 assumes that all the moments $\mathbf{f}_i^{(j_1,\dots,j_N)}$, in which $j_m > K$ at least for one $m = 1,\dots,N$, can be neglected. The number of unknowns and the number of equations in the system of linear Eq. (21) is $N(K+1)^N$.

4   The Steady-State Fluid Levels at Arbitrary Epoch

4.1   Equilibrium Relationships

Let $\widetilde{s}_i(\ell)$ the service time at station i in the ℓ-th cycle. The steady-state service time at station i and its mean is defined as

$$\widetilde{s}_i = \lim_{k \to \infty} \frac{\sum_{\ell=1}^{k} \widetilde{s}_i(\ell)}{k} \qquad \text{and} \qquad s_i = \lim_{k \to \infty} \frac{E[\sum_{\ell=1}^{k} \widetilde{s}_i(\ell)]}{k},$$

respectively. Similarly let $\widetilde{c}_i(\ell)$ the cycle time between two consecutive visit to station i in the ℓ-th cycle. The steady state cycle time at station i, and its mean is defined as

$$\widetilde{c}_i = \lim_{k \to \infty} \frac{\sum_{\ell=1}^{k} \widetilde{c}_i(\ell)}{k} \qquad \text{and} \qquad c_i = \lim_{k \to \infty} \frac{E[\sum_{\ell=1}^{k} \widetilde{c}_i(\ell)]}{k},$$

respectively. It follows from the definitions of c_i and s_i that

$$c_i = \sigma + \sum_{j=1}^{N} s_j, \qquad \text{and hence} \qquad c = c_i, \quad i \in \{1, \ldots, N\}. \tag{24}$$

Let $\Lambda_i(t)$ be the accumulated fluid flowed into the buffer of station i in interval $(0, t]$. The steady state mean amount of fluid, which flows into the buffer of station i during one cycle, a_i, is defined as

$$a_i = \lim_{k \to \infty} \frac{E[\sum_{\ell=1}^{k} \Lambda_i(t_i^f(\ell+1)) - \Lambda_i(t_i^f(\ell))]}{k}.$$

The right hand side of this definition can be rearranged as

$$\lim_{k \to \infty} \frac{E[\sum_{\ell=1}^{k} \Lambda_i(t_i^f(\ell+1)) - \Lambda_i(t_i^f(\ell))]}{E[\sum_{\ell=1}^{k} \widetilde{c}_i(\ell)]} \lim_{k \to \infty} \frac{E[\sum_{\ell=1}^{k} \widetilde{c}_i(\ell)]}{k}$$

and thus we get

$$a_i = \lambda_i c, \quad i \in \{1, \ldots, N\}. \tag{25}$$

Corollary 3. *In the stable fluid non-zero switchover-times polling model the steady-state mean cycle time can be expressed as*

$$c = \frac{\sigma}{1 - \rho}. \tag{26}$$

Proof. We apply a classical statistical equilibrium argumenting, see e.g. in [10]. The stable model is in statistical equilibrium, which implies that the mean amount of fluid flowing into the buffer of station i during a cycle equals the mean amount of fluid removed at station i during the same cycle, which equals $s_i d_i$. Putting them together yields

$$a_i = s_i d_i. \tag{27}$$

Applying (25) in (27) and expressing s_i from it leads to

$$s_i = \frac{\lambda_i}{d_i} c.$$ (28)

Applying (28) in (24) and changing to the notation of utilizations results in

$$c_i = \sigma + \sum_{j=1}^{N} \rho_j c.$$ (29)

Rearranging (29) gives the statement. □

Remark 3. The relations (24), (25) and (26) are valid independently of the used service discipline and hence they have more general validity scope.

4.2 The Steady-State Moments of the Service Time at Station i

The steady state pdf of the service time at station i, $s_i(t)$, and the corresponding LT, $s_i^*(v)$, for $t \geq 0$ are defined as

$$s_i(t) = \lim_{k \to \infty} \frac{d}{dt} \frac{E[\sum_{\ell=1}^{k} 1_{(\tilde{s}_i(\ell) < t)}]}{k}, \quad \text{and} \quad s_i^*(v) = \int_{t=0}^{\infty} s_i(t) e^{-st} dt,$$

where $1_{(\text{con})}$ denotes the indicator of condition "con".

Let $\mathbf{f}_i(x_i)$ and $\mathbf{f}_i^*(v)$ stand for steady-state vector density of the fluid level at station i at i-polling epoch and its LT, respectively. They can be obtained from $\mathbf{f}_i(\bar{x})$ and $\mathbf{f}_i^{(N)*}(\bar{v})$ as

$$\mathbf{f}_i(x_i) = \int_{x_1=0}^{\infty} \cdots \int_{x_{i-1}=0}^{\infty} \int_{x_{i+1}=0}^{\infty} \cdots \int_{x_N=0}^{\infty} \mathbf{f}_i(\bar{x}) \, dx_N \ldots dx_{i+1} dx_{i-1} \ldots dx_1,$$

$$\mathbf{f}_i^*(v) = \mathbf{f}_i^{(N)*}(\bar{v}) \Big|_{v_1 = \ldots = v_{i-1} = v_{i+1} = \ldots = v_N = 0, v_i = v}.$$

Theorem 2. *In the stable fluid non-zero switchover-times polling model with gated discipline the steady-state LT of the service time at station i can be expressed as*

$$s_i^*(v) = \mathbf{f}_i^*\left(\frac{v}{d_i}\right)\mathbf{e}, \quad i \in \{1, \ldots, N\}.$$ (30)

Proof. If the fluid level at station i is x_i at i-polling epoch then the service time at station i is $\frac{x_i}{d_i}$. Therefore the steady-state LT of the service time at station i can be obtained as

$$s_i^*(v) = \int_{x_i=0}^{\infty} \mathbf{f}_i(x_i) e^{-v\frac{x_i}{d_i}} dx_i \mathbf{e},$$ (31)

which can be rearranged as (30). □

Corollary 4. *In the stable fluid non-zero switchover-times polling model with gated discipline the steady-state moments of the service time at station i are given as*

$$s_i^{(k)} = \frac{1}{d_i^k} \mathbf{f}_i^{(k)} \mathbf{e}, \quad k \geq 1, \quad i \in \{1, \ldots, N\}. \tag{32}$$

Proof. Taking the k-th derivative of (30) with respect to v at $v = 0$ and multiplying it by $(-1)^k$ results in the statement. □

4.3 The Steady-State Joint Vector LT of the Fluid Levels at the Stations at Arbitrary Epoch

The steady-state joint density of the fluid levels at the stations and the state of the modulating Markov chain at an arbitrary epoch, the $1 \times L$ row vector $\mathbf{q}(\bar{x})$ is defined by its j-th element as

$$[\mathbf{q}(\bar{x})]_j = \lim_{t \to \infty} \frac{\partial}{\partial x_1} \cdots \frac{\partial}{\partial x_N} Pr(\Omega(t) = j, X_1(t) < x_1, \ldots X_N(t) < x_N), \ j \in \Omega,$$

and its LT with respect to $\bar{x}$ can be given as

$$\mathbf{q}^{(N)*}(\bar{v}) = \int_{x_1=0}^{\infty} \cdots \int_{x_N=0}^{\infty} \mathbf{q}(\bar{x}) e^{-v_1 x_1} \ldots e^{-v_N x_N} dx_N \ldots dx_1.$$

Moreover, let $\mathbf{e}_j = (0, \ldots, 0, 1, 0, \ldots, 0)$ be the $1 \times L$ vector with 1 at the j-th position. Then the $1 \times L$ indicator vector $\mathbf{1}_{(\Omega(t))}$ is defined as

$$\mathbf{1}_{(\Omega(t))} = \sum_{j=1}^{L} \mathbf{1}_{(\Omega(t)=j)} \mathbf{e}_j.$$

We use the following notation

$$\mathbf{f}_i^{(N-1)*}(v_1, \ldots, v_{i-1}, x_i, v_{i+1}, \ldots, v_N) = \int_{x_1=0}^{\infty} \cdots \int_{x_{i-1}=0}^{\infty} \int_{x_{i+1}=0}^{\infty} \cdots \int_{x_N=0}^{\infty}$$
$$\mathbf{f}_i(\bar{x}) \, e^{-v_1 x_1} \ldots e^{-v_{i-1} x_{i-1}} e^{-v_{i+1} x_{i+1}} \ldots e^{-v_N x_N} dx_N \ldots dx_{i+1} dx_{i-1} \ldots dx_1.$$

Theorem 3. *In the stable fluid non-zero switchover-times polling model with gated discipline the following relation holds for the steady-state joint vector LT of the fluid levels at the stations at arbitrary epoch:*

$$\mathbf{q}^{(N)*}(\bar{v}) \left(\sum_{j=1}^{N} \mathbf{R}_j v_j - \mathbf{Q} \right) = \tag{33}$$

$$\frac{1}{c} \sum_{i=1}^{N} \left[d_i v_i \left(\mathbf{f}_i^{(N)*}(\bar{v}) - \mathbf{m}_i^{(N)*}(\bar{v}) \right) \left(\sum_{j \neq i} \mathbf{R}_j v_j + (\mathbf{R}_i - d_i \mathbf{I}) v_i - \mathbf{Q} \right)^{-1} \right].$$

Proof. The fluid levels at the stations at arbitrary epoch can be expressed by the help of the fluid levels at the last i-polling epoch on LT level by utilizing the transient behavior of the arrived fluid (relation (6)) and taking into account that it can fall either in service or switchover period as well as its position in the actual period. Thus it is enough to average over a polling cycle for determining the behavior at arbitrary epoch.

Therefore $\mathbf{q}^{(N)*}(\bar{v})$ is given by

$$
\mathbf{q}^{(N)*}(\bar{v}) = \frac{E[\int_{t=0}^{\tilde{c}_1} e^{-\sum_{j=1}^{N} X_j(t)v_j} \mathbf{1}_{(\Omega(t))} dt]}{E[\tilde{c}_1]} \tag{34}
$$

$$
= \frac{\sum_{i=1}^{N} E[\int_{t=0}^{\tilde{s}_i} e^{-\sum_{j=1}^{N} X_j(t)v_j} \mathbf{1}_{(\Omega(t))} dt] + \sum_{i=1}^{N} E[\int_{t=0}^{\tilde{\sigma}_i} e^{-\sum_{j=1}^{N} X_j(t)v_j} \mathbf{1}_{(\Omega(t))} dt]}{c}.
$$

The fluid level at time t at station i in the service time of station i is the sum of the remaining fluid level, $\xi - td_i$, and the fluid level arrived during t. The fluid level at time t at other stations, i.e., $j \neq i$ in the service time of station i is the sum of the fluid level at the begin of the service time and the fluid amount arrived during t.

Taking into account the state change of the modulating CTMC from 0 to t the LT term $E[\int_{t=0}^{\tilde{s}_i} e^{-\sum_{j=1}^{N} X_j(t)v_j} \mathbf{1}_{(\Omega(t))} dt]$ can be given as

$$
E[\int_{t=0}^{\tilde{s}_i} e^{-\sum_{j=1}^{N} X_j(t)v_j} \mathbf{1}_{(\Omega(t))} dt] \tag{35}
$$

$$
= \int_{\xi=0}^{\infty} e^{-(\xi-td_i)v_i} \mathbf{f}_i^{(N-1)*}(v_1, \ldots, v_{i-1}, \xi, v_{i+1}, \ldots, v_N) \int_{t=0}^{\frac{\xi}{d_i}} \mathbf{A}^{(N)*}(t, \bar{v}) dt d\xi
$$

$$
= \int_{\xi=0}^{\infty} e^{-\xi v_i} \mathbf{f}_i^{(N-1)*}(v_1, \ldots, v_{i-1}, \xi, v_{i+1}, \ldots, v_N) \int_{t=0}^{\frac{\xi}{d_i}} e^{td_i v_i} \mathbf{A}^{(N)*}(t, \bar{v}) dt d\xi.
$$

Applying (6) in (35) and rearrangement gives

$$
E[\int_{t=0}^{\tilde{s}_i} e^{-\sum_{j=1}^{N} X_j(t)v_j} \mathbf{1}_{(\Omega(t))} dt] = \int_{\xi=0}^{\infty} e^{-\xi v_i} \mathbf{f}_i^{(N-1)*}(v_1, \ldots, v_{i-1}, \xi, v_{i+1}, \ldots, v_N)
$$

$$
\int_{t=0}^{\frac{\xi}{d_i}} e^{-t(\sum_{j\neq i} \mathbf{R}_j v_j + (\mathbf{R}_i - d_i \mathbf{I})v_i - \mathbf{Q})} dt d\xi. \tag{36}
$$

The internal integral can be evaluated by means of a relation, which can be obtained by the help of the Taylor-expansion of $e^{\mathbf{Z}t}$, and is given by

$$
\int_{t=0}^{x} e^{-\mathbf{Z}t} dt \mathbf{Z} = (\mathbf{I} - e^{-\mathbf{Z}x}). \tag{37}
$$

Applying (37) in (36) and rearrangement yields

$$E[\int_{t=0}^{\widetilde{s}_i} e^{-\sum_{j=1}^{N} X_j(t)v_j} \mathbf{1}_{(\Omega(t))} dt] \left(\sum_{j \neq i} \mathbf{R}_j v_j + (\mathbf{R}_i - d_i \mathbf{I}) v_i - \mathbf{Q} \right) \quad (38)$$

$$= \int_{\xi=0}^{\infty} e^{-\xi v_i} \mathbf{f}_i^{(N-1)*}(v_1, \ldots, v_{i-1}, \xi, v_{i+1}, \ldots, v_N)$$

$$\left(\mathbf{I} - e^{-\frac{\xi}{d_i} \left(\sum_{j \neq i} \mathbf{R}_j v_j + (\mathbf{R}_i - d_i \mathbf{I}) v_i - \mathbf{Q} \right)} \right) d\xi.$$

Rearrangement and applying (13) in (38) leads to

$$E[\int_{t=0}^{\widetilde{s}_i} e^{-\sum_{j=1}^{N} X_j(t)v_j} \mathbf{1}_{(\Omega(t))} dt] \left(\sum_{j \neq i} \mathbf{R}_j v_j + (\mathbf{R}_i - d_i \mathbf{I}) v_i - \mathbf{Q} \right) \quad (39)$$

$$= \mathbf{f}_i^{(N)*}(\bar{v}) - \mathbf{f}_i^{(N)*}(v_1, \ldots, v_{i-1}, \frac{\sum_{i=1}^{N} \mathbf{R}_i v_i - \mathbf{Q}}{d_i}, v_{i+1}, \ldots, v_N)$$

$$= \mathbf{f}_i^{(N)*}(\bar{v}) - \mathbf{m}_i^{(N)*}(\bar{v}).$$

Further rearranging of (39) yields

$$E[\int_{t=0}^{\widetilde{s}_i} e^{-\sum_{j=1}^{N} X_j(t)v_j} \mathbf{1}_{(\Omega(t))} dt] \left(\sum_{j=1}^{N} \mathbf{R}_j v_j - \mathbf{Q} \right) \quad (40)$$

$$= \mathbf{f}_i^{(N)*}(\bar{v}) - \mathbf{m}_i^{(N)*}(\bar{v}) + d_i v_i E[\int_{t=0}^{\widetilde{s}_i} e^{-\sum_{j=1}^{N} X_j(t)v_j} \mathbf{1}_{(\Omega(t))} dt].$$

Now we consider the term $E[\int_{t=0}^{\widetilde{\sigma}_i} e^{-\sum_{j=1}^{N} X_j(t)v_j} \mathbf{1}_{(\Omega(t))} dt]$. The fluid level at time t at station j, $j \in \{1, \ldots, N\}$, in the switchover time after the service of station i is the sum of the fluid level at station j at start of the switchover time, and the fluid level arrived during t. Taking into account the state change of the modulating CTMC from 0 to t the LT term $E[\int_{t=0}^{\widetilde{\sigma}_i} e^{-\sum_{j=1}^{N} X_j(t)v_j} \mathbf{1}_{(\Omega(t))} dt]$ can be given as

$$E[\int_{t=0}^{\widetilde{\sigma}_i} e^{-\sum_{j=1}^{N} X_j(t)v_j} \mathbf{1}_{(\Omega(t))} dt] = \mathbf{m}_i^{(N)*}(\bar{v}) \int_{\tau=0}^{\infty} \int_{t=0}^{\tau} \mathbf{A}^{(N)*}(t, \bar{v}) dt \, \sigma(\tau) \, d\tau. \quad (41)$$

Applying (6) in (41) yields

$$E[\int_{t=0}^{\widetilde{\sigma}_i} e^{-\sum_{j=1}^{N} X_j(t)v_j} \mathbf{1}_{(\Omega(t))} dt]$$

$$= \mathbf{m}_i^{(N)*}(\bar{v}) \int_{\tau=0}^{\infty} \int_{t=0}^{\tau} e^{-t(\sum_{j=1}^{N} \mathbf{R}_j v_j - \mathbf{Q})} dt \, \sigma(\tau) \, d\tau. \quad (42)$$

We apply again (37), now in (42), which gives

$$E[\int_{t=0}^{\tilde{\sigma}_i} e^{-\sum_{j=1}^{N} X_j(t)v_j} \mathbf{1}_{(\Omega(t))} dt] \left(\sum_{j=1}^{N} \mathbf{R}_j v_j - \mathbf{Q}\right) = \tag{43}$$

$$\mathbf{m}_i^{(N)*}(\bar{v}) \int_{\tau=0}^{\infty} \left(\mathbf{I} - e^{-\tau(\sum_{j=1}^{N} \mathbf{R}_j v_j - \mathbf{Q})}\right) \sigma(\tau)\, d\tau.$$

Rearranging (42) and applying (14) in it gives the relation for $E[\int_{t=0}^{\tilde{\sigma}_i} e^{-\sum_{j=1}^{N} X_j(t)v_j} \mathbf{1}_{(\Omega(t))} dt]$ as

$$E[\int_{t=0}^{\tilde{\sigma}_i} e^{-\sum_{j=1}^{N} X_j(t)v_j} \mathbf{1}_{(\Omega(t))} dt] \left(\sum_{j=1}^{N} \mathbf{R}_j v_j - \mathbf{Q}\right) \tag{44}$$

$$= \mathbf{m}_i^{(N)*}(\bar{v}) \left(\mathbf{I} - \sigma_i^* \left(\sum_{j=1}^{N} \mathbf{R}_j v_j - \mathbf{Q}\right)\right) = \mathbf{m}_i^{(N)*}(\bar{v}) - \mathbf{f}_{i+1}^{(N)*}(\bar{v}).$$

Using (40) and (44) in (34) and rearranging gives

$$\mathbf{q}^{(N)*}(\bar{v}) \left(\sum_{j=1}^{N} \mathbf{R}_j v_j - \mathbf{Q}\right)$$

$$= \frac{1}{c} \left(\sum_{i=1}^{N} \left(\mathbf{f}_i^{(N)*}(\bar{v}) - \mathbf{m}_i^{(N)*}(\bar{v}) + d_i v_i E[\int_{t=0}^{\tilde{s}_i} e^{-\sum_{j=1}^{N} X_j(t)v_j} \mathbf{1}_{(\Omega(t))} dt]\right)\right.$$

$$\left. + \sum_{i=1}^{N} \left(\mathbf{m}_i^{(N)*}(\bar{v}) - \mathbf{f}_{i+1}^{(N)*}(\bar{v})\right)\right)$$

$$= \frac{1}{c} \sum_{i=1}^{N} d_i v_i E[\int_{t=0}^{\tilde{s}_i} e^{-\sum_{j=1}^{N} X_j(t)v_j} \mathbf{1}_{(\Omega(t))} dt]. \tag{45}$$

The statement of the theorem comes by applying (39) in (45).   □

Let $\mathbf{q}_i^*(v)$ denote the steady-state vector LT of the fluid level at station i at arbitrary epoch. $\mathbf{q}_i^*(v)$ can be obtained as

$$\mathbf{q}_i^*(v) = \mathbf{q}^{(N)*}(\bar{v})\Big|_{v_1=\ldots=v_{i-1}=v_{i+1}=\ldots=v_N=0, v_i=v}.$$

Let $\mathbf{m}_i(x_i)$ and $\mathbf{m}_i^*(v)$ stand for steady-state vector density of the fluid level at station i at i-departure epoch and its LT, respectively. They can be obtained from $\mathbf{m}_i(\bar{x})$ and $\mathbf{m}_i^{(N)*}(\bar{v})$ as

$$\mathbf{m}_i(x_i) = \int_{x_1=0}^{\infty} \ldots \int_{x_{i-1}=0}^{\infty} \int_{x_{i+1}=0}^{\infty} \ldots \int_{x_N=0}^{\infty} \mathbf{m}_i(\bar{x})\, dx_N \ldots dx_{i+1} dx_{i-1} \ldots dx_1,$$

$$\mathbf{m}_i^*(v) = \mathbf{m}_i^{(N)*}(\bar{v})\Big|_{v_1=\ldots=v_{i-1}=v_{i+1}=\ldots=v_N=0, v_i=v}.$$

Corollary 5. *In the stable fluid non-zero switchover-times polling model with gated discipline the following relation holds for the steady-state vector LT of the fluid level at station i at arbitrary epoch:*

$$\mathbf{q}_i^*(v)\,(\mathbf{R}_i v - \mathbf{Q})\,((\mathbf{R}_i - d_i\mathbf{I})\,v - \mathbf{Q}) = \frac{1}{c}d_i v\,(\mathbf{f}_i^*(v) - \mathbf{m}_i^*(v))\,. \tag{46}$$

Proof. The statement comes by setting $v_1 = \ldots = v_{i-1} = v_{i+1} = \ldots = v_N = 0, v_i = v$ in (33). $\qquad\square$

Remark 4. The relation (46) holds also for fluid vacation model with gated discipline (see (61) in [6]).

Corollary 6.   *In the stable fluid non-zero switchover-times polling model with gated discipline the steady-state vector mean of the fluid level at station i at arbitrary epoch can be determined as*

$$
\begin{aligned}
\mathbf{q}_i^{(1)} &= \frac{1}{6\lambda_i(\lambda_i - d_i)}\mathbf{r}^{(3)}\mathbf{e}\boldsymbol{\pi} \\
&\quad - \frac{1}{2(\lambda_i - d_i)}\mathbf{r}^{(2)}\frac{1}{\lambda_i}\left(\mathbf{I} - \frac{1}{(\lambda_i - d_i)}\mathbf{e}\boldsymbol{\pi}(\mathbf{R}_i - d_i\mathbf{I})\right) \\
&\quad \times (\mathbf{Q} + \mathbf{e}\boldsymbol{\pi})^{-1}(\mathbf{R}_i - d_i\mathbf{I})\mathbf{e}\boldsymbol{\pi} \\
&\quad - \frac{1}{2(\lambda_i - d_i)}\mathbf{r}^{(2)}\mathbf{e}\boldsymbol{\pi}\,(\mathbf{Q} + \mathbf{e}\boldsymbol{\pi})^{-1}\left(\frac{\mathbf{R}_i\mathbf{e}\boldsymbol{\pi}}{\lambda_i} - \mathbf{I}\right) \\
&\quad + \mathbf{r}^{(1)}\,(\mathbf{Q} + \mathbf{e}\boldsymbol{\pi})^{-1}\left(\frac{1}{(\lambda_i - d_i)}(\mathbf{R}_i - d_i\mathbf{I})\mathbf{e}\boldsymbol{\pi} - \mathbf{I}\right) \\
&\quad \times \left(\frac{-1}{\lambda_i(\lambda_i - d_i)}(\mathbf{R}_i - d_i\mathbf{I})\,(\mathbf{Q} + \mathbf{e}\boldsymbol{\pi})^{-1}(\mathbf{R}_i - d_i\mathbf{I})\mathbf{e}\boldsymbol{\pi}\right. \\
&\quad \left. + (\mathbf{Q} + \mathbf{e}\boldsymbol{\pi})^{-1}\left(\frac{\mathbf{R}_i\mathbf{e}\boldsymbol{\pi}}{\lambda_i} - \mathbf{I}\right)\right) \\
&\quad + \boldsymbol{\pi}\mathbf{R}_i\,(\mathbf{Q} + \mathbf{e}\boldsymbol{\pi})^{-1}\left(\frac{\mathbf{R}_i\mathbf{e}\boldsymbol{\pi}}{\lambda_i} - \mathbf{I}\right).
\end{aligned} \tag{47}
$$

where c is given by (26) and $\mathbf{r}^{(1)}$, $\mathbf{r}^{(2)}$ *and* $\mathbf{r}^{(3)}$ *are given by*

$$
\begin{aligned}
\mathbf{r}^{(1)} &= -\frac{d_i}{c}(\mathbf{f} - \mathbf{m}), \\
\mathbf{r}^{(2)} &= -\frac{2d_i}{c}(\mathbf{f}^{(1)} - \mathbf{m}^{(1)}), \\
\mathbf{r}^{(3)} &= -\frac{3d_i}{c}(\mathbf{f}^{(2)} - \mathbf{m}^{(2)}).
\end{aligned}
$$

Proof. The proof of the statement can be found in [6] (proof of Corollary 6). $\square$

Acknowledgment. This work is supported by the OTKA K-123914 project and by the ÚNKP-17-4-III New National Excellence Program of the Ministry of Human Capacities.

References

1. Dai, J.G.: On the positive Harris recurrence for multiclass queueing networks: a unified approach via fluid limit models. Ann. Appl. Prob. **5**, 49–77 (1995)
2. Dai, J.G., Meyn, S.P.: Stability and convergence of moments for multiclass queueing networks via fluid limit models. IEEE Trans. Automat. Control **40**, 1–16 (1995)
3. Kulkarni, V.G.: Fluid models for single buffer systems. In: Dshalalow, J. (ed.) Frontiers in Queueing, pp. 321–338. CRC, Boca Raton (1997)
4. Ahn, S., Ramaswami, V.: Efficient algorithms for transient analysis of stochastic fluid flow models. J. Appl. Probab. **42**(2), 531–549 (2005)
5. Czerniak, O., Yechiali, U.: Fluid polling systems. Queueing Syst. **63**, 401–435 (2009)
6. Saffer, Z., Telek, M.: Fluid vacation model with Markov modulated load and gated discipline. In: 9th International Conference on Queueing Theory and Network Applications (QTNA2014), pp. 184–197 (2014)
7. Saffer, Z., Telek, M.: Fluid vacation model with markov modulated load and exhaustive discipline. In: Horváth, A., Wolter, K. (eds.) EPEW 2014. LNCS, vol. 8721, pp. 59–73. Springer, Cham (2014). https://doi.org/10.1007/978-3-319-10885-8_5
8. Saffer, Z., Telek, M.: Exhaustive fluid vacation model with Markov modulated load. Perform. Eval. (PEVA) **98**, 19–35 (2016)
9. Horváth, G., Telek, M.: Exhaustive fluid vacation model with positive fluid rate during service. Perform. Eval. **91**, 286–302 (2015)
10. Eisenberg, M.: Queues with periodic service and changeover time. Oper. Res. **20**, 440–451 (1972)

Modelling Large Timescale and Small Timescale Service Variability

Marco Gribaudo[1], Illés Horváth[2(✉)], Daniele Manini[3], and Miklós Telek[4]

[1] Dip. di Elettronica, Informazione e Bioingengeria, Politecnico di Milano,
Milan, Italy
`marco.gribaudo@polimi.it`
[2] MTA-BME Information Systems Reseach Group, Budapest, Hungary
`horvath.illes.antal@gmail.com`
[3] Dip. di Informatica, Università di Torino, Turin, Italy
`manini@di.unito.it`
[4] Dept. of Networked systems, Technical University of Budapest, Budapest, Hungary
`telek@hit.bme.hu`

Abstract. The performance of service units might depend on various randomly changing environmental effects. It is quite often the case that these effects varies on different time scales. In this paper we consider short and long scale service variability, where the short scale variability affects the instantaneous service speed of the service unit and the large scale effect is defined by a modulating background Markov chain. The main modelling challenge is that the considered short and long range variation results randomness along different axes, the short scale variability along the time axis and the long scale variability along the work axis. The work presents mostly simulation results; the mathematical setup for analytical results is provided, but the actual analysis is subject to future research.

Keywords: Short and long term service variability · Brownian motion
Markov modulation · Performance analysis

1 Introduction

Service speed variability is a problem that has been measured in many practical application scenario. For example in [3], it has been observed for vehicular traffic. More recently this problem has been recognized in data-center [2]. The effect of variability was also studied in [1] with application to video-streaming. Most of the previous literature however, focused only on large-time scale variability, where Markov-modulating models represent the random effect of the environment. All of those models can be handled with matrix analytic methods, summarized e.g., by Latouche and Ramaswami in [4].

This work is partially supported by the OTKA K-123914 grant.

The variation in the service speed can be modelled by dividing the amount of job to be executed into "infinitesimal quantities of work to be done" and consider the "speed at which this infinitesimal work is performed", i.e., the random amount of time needed to execute the infinitesimal amount of work. Then, if a model that defines how speed changes over time, the complete system can be modelled in a straight-forward way where the amount of work increases gradually along the analysis and the time required to execute the given amount of work is a random process.

If the service process depends on a time dependent random process, e.g., on a modulating background CTMC representing the environmental state, whose "clock" evolves according to the time, then the natural performance analysis is based on the gradually increasing time and randomly varying time dependent environment state.

However, in many real applications, variability is not easily predictable and works at different time-scales. Modulating CTMCs (whose "clock" evolves according to the time) works very well to model variability where the parameters of the job execution remains constant for a longer random period of time, and there are few jumps during the execution of one job. Apart of this large scale variability, in this work, we focus also on variability that occurs at much smaller time scales, where the execution speeds changes thousands, if not millions, of times during the execution of the main job, and combine it with the more classical modulation that works on a larger time scale.

The remainder of this paper is structured as follows. In Sect. 2 we start with considering only the small time-scale variability. In Sect. 3 we additionally introduce also the large time-scale variability. The effects of the considered variability is studied in Sect. 4 through numerical examples, and Sect. 5 concludes the paper.

2   Small Time-Scale Variability

In this section, we omit the large time-scale variability and instead focus only on small time-scale variability. So assume that the environmental state is unchanged for now.

We introduce a second order fluid model for the short time-scale variability: assuming that a job is composed of quantums of size Δx, each such quantum is served in a random amount of time with distribution $N(\mu\Delta x, \sigma^2\Delta x)$ (with $\mu > 0$). Assuming that the service times of the different quantums are independent, the progress of service is modeled by a Brownian motion $B(x)$ with parameters μ and σ^2. We emphasize that in this model, the Brownian motion corresponds to *the time required to service a job as a function of the size of the job* (see Fig. 1). A job of size x thus requires a random time T with distribution $N(\mu x, \sigma^2 x)$, whose probability density function is $\frac{e^{-\frac{(t-\mu x)^2}{2x\sigma^2}}}{\sqrt{2\pi x\sigma^2}}$. Note that a Brownian motion may take negative values as well, which does not make sense physically, but, since $\mu > 0$, for macroscopic values of w, the probability that T is negative is negligible.

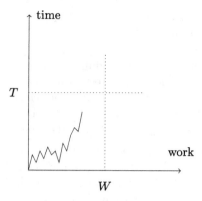

Fig. 1. The time T required to service a job as a function of the job size W

We focus on the service of a job in a queue whose work requirement, W, is generally distributed according to probability density function $f_W(x)$.

Using the second order fluid model assumption, the probability density function of the service time of a job, denoted by $f_T(t)$, can be computed as:

$$f_T(t) = \int_0^\infty f_W(x) \cdot \frac{e^{-\frac{(t-\mu x)^2}{2x\sigma^2}}}{\sqrt{2\pi x\sigma^2}} dx \tag{1}$$

we can compute the moments of T as:

$$
\begin{aligned}
E[T^K] &= \int_0^\infty f_W(x) \sum_{j=0}^k u_{k,j}(x\mu)^j (\sqrt{x}\sigma)^{k-j} dx = \\
&= \sum_{j=0}^k u_{k,j}\mu^j \sigma^{k-j} \int_0^\infty f_W(x) x^j (\sqrt{x})^{k-j} dx = \\
&= \sum_{j=0}^k u_{k,j}\mu^j \sigma^{k-j} E[W^{\frac{k+j}{2}}]
\end{aligned}
\tag{2}
$$

Since σ^2 appears only for even exponents, $k+i$ is always even, so $E[T^{\frac{k+i}{2}}]$ is always an integer moment of T. For example, for the first and second moment, since $E[N(x\mu, x\sigma^2)^2] = x\mu$ and $E[N(x\mu, x\sigma^2)^2] = x^2\mu^2 + x\sigma^2$, we have:

$$E[T] = \mu E[W],$$
$$E[T^2] = \mu^2 E[W^2] + \sigma^2 E[W].$$

3   Combining Large and Small Time-Scale Variability

Large scale variability can be considered using a discrete state Markov modulating process (MMP) of K states, denoted by $M(t)$. The MMP is a CTMC with infinitesimal generator matrix denoted by Q. In state i, the service is characterised by rate μ_i and variance σ_i.

Only considering large scale variability (that is, assuming $\sigma_k \equiv 0$) would lead to a standard first order Markov-modulated fluid model. However, including small-scale variability makes for an interesting and complex model.

Assume that a job of size $W = x$ starts service at time $t = 0$, with the background modulating process in state i. Then the evolution of the service time $B(x)$ as a function of the job size is the following:

- Let a_1 denote the time of the first transition of $M(t)$. As long as $B(x)$ is smaller than a_1, $B(x)$ evolves according to a $\mathrm{BM}(\mu_i, \sigma_i)$.
- At time a_1, $M(t)$ changes to some state j. Accordingly, assuming that the first passage of $B(x)$ to a_1 occurs at work amount w_1, for $x \geq w_1$, $B(x)$ evolves according to a $\mathrm{BM}(\mu_j, \sigma_j)$ (starting from the point w_1 and from level a_1).
- This is repeated for further transitions of $M(t)$ at times $a_2, a_3, \ldots$, up to the point $x = W$.

Note that in visualization, the x axis denotes the job size, and the y axis denotes time, see Fig. 2. Thus for $B(x)$, the behaviour can be described as a type of *level-dependent Brownian motion*: the parameters μ and σ of the Brownian motion change upon first passage to levels $a_1, a_2, \ldots$. This is different from usual second order Markov-modulated fluid models, where parameter changes occur upon the variable of the Brownian motion (x in our case) reaching some transition points instead of the level reaching transition points.

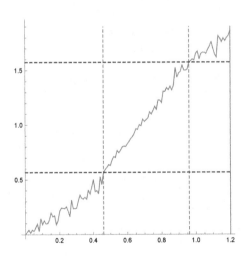

Fig. 2. A possible realization of $B(x)$ for job size $W = 1.2$

Keeping in mind that $M(t)$ is a CTMC, the entire distribution of $B(x)$ is determined by the initial points $t = 0$ and $x = 0$ and the initial state of the modulating process $M(0) = i$. The process $B(x)$ can be simulated as follows:

- $B(x)$ starts from $t = 0$, $x = 0$, with $M(0) = i$ and job size W.
- Generate the first transition time a_1 of $M(t)$.
- $B(x)$ runs as a BM(μ_i, σ_i^2) until either the value of $B(x)$ reaches a_1 or x reaches W, whichever occurs first.
- If $x = W$ occurred first, then the simulation is finished.
- If $B(w_1) = a_1$ for some $w_1 < W$, then we generate the next state j and also the next transition time a_2 according to the CTMC $M(t)$, then continue $B(x)$ as a Brownian motion with parameters (μ_j, σ_j^2) starting from the point (w_1, a_1) until either the value of $B(x)$ reaches a_2 or x reaches W, whichever occurs first.
- We keep generating new transitions and new Brownian motion sections until we reach W. The service time of the job is $T = B(W)$.

The main question, similar to Sect. 2, is the distribution of T and performance measures derived from T. In this case, an analytical answer is non-trivial even for a given job size $W = x$. One possible analytic formulation is to first introduce the cumulative distribution type functions (for fixed x)

$$G_{ij}(x,t) = \Pr\left(B(x) \leq t, M(B(x)) = j | M(0) = i, W = x\right) \qquad (3)$$

which include information about the initial and final background state of $M(t)$ along with the distribution of the service time. An analytic formula for $G_{ij}(W,t)$ is subject to ongoing research.

4 Simulation Results

To study the effects of variability, we have applied the procedure outlined in Sect. 3 to simulate the behaviour of the queue with short and long scale variability. In particular, to find the intersection between the Brownian motion and the level determined by the time at which the modulating process changes state, we have discretised the work with a quantum Δx, and during the period when the MMP stays in state i, for each quantum we have set the evolution of the time according to a normal distribution $N(\mu_i \Delta x, \sigma_i^2 \Delta x)$ (following the procedure outlined at the beginning of Sect. 2). The MMP leaves state i at the first time instant in which the discretised BM crosses the level T_n, where T_n is the time of the nth state transition of the MMP. When the nth state transition occurs in state i, then $T_n = T_{n-1} + \tau_i$, where T_{n-1} is the time of the previous state transition and τ_i is exponentially distributed with parameter $-Q_{ii}$ (the $i$the diagonal element of the generator matrix of the modulating CTMC). This simulation approach is indeed an approximation, but it can be made arbitrarily precise by choosing appropriately small values of Δx.

In our numerical experiment, we have considered a two-state modulating process with jump rates γ_{12} and γ_{21}, and studied the effects of different service speed and variability parameters μ_i and σ_i. To show a possible application, we have used the proposed process to describe the variable service rate in an M/G/1 queue, where jobs arrive according to a Poisson process of rate λ and are served by a single server subject to short and long range variability according to a first-come-first-served discipline. To compare the results for different service time distributions we assumed that the mean service time $E[W]$ is identical in each cases. The arrival rate, λ, is selected such that the queue is stable. Unless otherwise stated, the used parameters have been the following:

$$\lambda = \tfrac{1000}{350}\text{job/s}, \quad E[W] = 100\,\text{ms}, \quad \Delta X = 0.05\,\text{ms},$$
$$\mu_1 = 2, \quad \mu_2 = 4, \quad \sigma_1 = 0.4, \quad \sigma_2 = 1.5, \tag{4}$$
$$\tfrac{1}{\gamma_{12}} = 1.25\,\text{s}, \quad \tfrac{1}{\gamma_{21}} = 0.8\,\text{s}.$$

In this framework, the discretisation interval has been chosen so that on average, the BM for each job requires 2000 samples, and in the average sojourn time in the two modulating states, the BM is samples respectively 25000 and 16000 times. Each simulation considers the execution of $N = 10000$ jobs.

We start focusing on jobs requiring a fixed amount of work (i.e. $W = E[W]$ is deterministic. Figure 3a shows the service time distribution for different server variability configuration. The *Base* case, considers the case in which no variability is used: in particular to $\mu_1 = \mu_2 = 2.4848$ and $\sigma_1 = \sigma_2 = 0$. As it is expected, all the probability mass is centred along $\mu E[W] = 248.48$. The *Small* variability cases, differs from the *Base* one by adding a small variability. In the *Small (fixed)* case $\sigma_1 = \sigma_2 = 0.98773$ and in the *Small (variable)* case we have the state dependent variability $\sigma_1 = 0.4$ and $\sigma_2 = 1.5$. As it can be seen, they both destroy the deterministic behaviour, in a slightly different way: the fixed σ case has a more uniform effect, while the variable one presents larger tails. The case called *Large* considers only large scale variability only, i.e., $\sigma_1 = \sigma_2 = 0$. During a sojourn in a state of the MMP the service time of a job is deterministic. In state 1, with $\mu_1 = 2$, the service time is exactly 200 ms, and in state 2, with $\mu_2 = 4$, it is exactly 400 ms. The jumps in Fig. 3a at 200 ms, and 400 ms are associated with the cases when the MMP stays in state 1 (2, respectively) for the whole period of the service. The cases when the MMP experiences state transition during the service are represented by the continuously increasing part of the *Large* curve. The case that combines both small and large scale variability (*Small+Large*, $\mu_1 = 2, \mu_2 = 4, \sigma_1 = 0.4, \sigma_2 = 1.5$) further smooths the curves, and the effect is more evident near the two probability masses at 200 ms, and 400 ms. Figure 3b shows the response time distribution of the corresponding queuing models. In this case it is interesting to see that in the cases where small variability is considered there are no jumps due to its perturbation effect.

We then study the effect of the modulating process, by changing the average sojourn time in its two states, while maintaining the state probabilities. Figure 4 considers different combinations of sojourn times ranging from 12.5 s and 8 s down to 1.25 ms and 0.8 ms for the deterministic job length distribution W, and the other parameters defined as in (4). When the sojourn time is very large, service times are correlated, and the service time distribution tend to concentrate the probability mass near the times required in both modulating states. On the other hand, when the switching process changes very fast, the distribution tend to concentrate in the average case, producing results very similar to the one seen

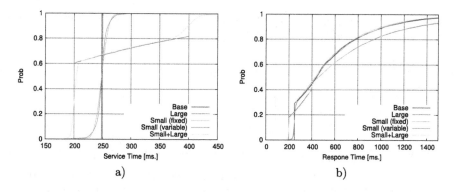

Fig. 3. Considering different small scale and large scale variability configurations for a fixed job length: (a) service time distribution, (b) response time distribution

in Fig. 3 for the cases with small variability only: in this case, there is almost no difference between large scale and small scale variability, because the quick alternation of the modulating process eliminates the large scale effect. As a final remark, in order to consider a switching process with 1.25 ms and 0.8 ms, we had to reduce the sampling time $\Delta x = 0.01$ to allow a sufficient number of samples during the sojourn in a modulating state. For what concerns response time (Fig. 4c), when the modulating process present deep correlation by spending longer times in a single state, bursts are created, decreasing considerably the performances of the system.

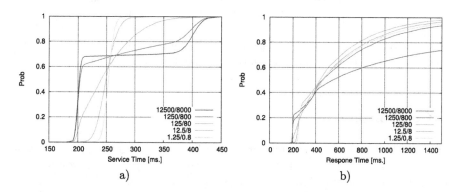

Fig. 4. Considering different durations in the modulating process for a fixed job length: (a) service time distribution, (b) response time distribution

We finally consider the effect of variability on different job length distributions. In particular, Fig. 5a shows the service time distribution when the job length follows, beside the deterministic distribution already discussed, an exponential distribution, an Erlang distribution with 4 stages, and the following

Hyper-Exponential $(w_H(x))$ and Pareto $(w_P(x))$ distributions characterised by the following probability density functions:

$$w_H(x) = \frac{1}{2}\lambda_1 e^{-\lambda_1 x} + \frac{1}{2}\lambda_2 e^{-\lambda_2 x},$$

$$w_P(x) = \begin{cases} \dfrac{20^{\frac{5}{4}}\frac{5}{4}}{x^{\frac{9}{4}}} & x > 20, \\ 0 & x < 20, \end{cases}$$

where $\lambda_1 = \dfrac{1}{100(1+\sqrt{\frac{3}{5}})}$ and $\lambda_2 = \dfrac{1}{100(1-\sqrt{\frac{3}{5}})}$. As it can be noted, the effect of service variability is more evident on job length distributions with a lower coefficient of variation. Figure 5b shows the effect on response time: indeed, combining the effect of service variability with heavy tailed distribution, as for the Pareto case, can create very long queues which can lead to extremely long response times.

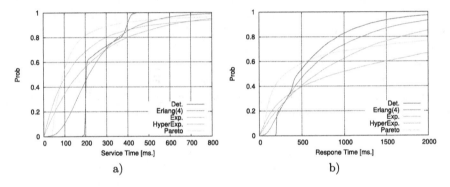

Fig. 5. Considering small scale and large scale variability for different job length distributions: (a) service time distribution, (b) response time distribution

5   Conclusions

In this work, we have introduced a queue with a service model where the large timescale variability is modelled by a modulating background Markov process, and small timescale variability is modelled by a second-order fluid process for the service time of a job. The resulting service model can be interpreted as a certain type of level-dependent Brownian motion.

We have presented simulation results for the service time and response time of a job for various job size distributions. In future work, we hope to give a full analytic description of the system, most notably by giving an analytic solution for (3).

References

1. Anjum, B., Perros, H.: Bandwidth estimation for video streaming under percentile delay, jitter, and packet loss rate constraints using traces. Comput. Commun. **57**, 73–84 (2015). http://www.sciencedirect.com/science/article/pii/S0140366414003089
2. Guo, J., Liu, F., Huang, X., Lui, J.C., Hu, M., Gao, Q., Jin, H.: On efficient bandwidth allocation for traffic variability in datacenters, April 2014
3. Kimber, R., Daly, P.: Time-dependent queueing at road junctions: observation and prediction. Transp. Res. Part B Methodol. **20**(3), 187–203 (1986). http://www.sciencedirect.com/science/article/pii/0191261586900160
4. Latouche, G., Ramaswami, V.: Introduction to matrix analytic methods in stochastic modeling. Soc. Ind. Appl. Math. (1999). https://doi.org/10.1137/1.9780898719734

Queueing Models for Cloud Computing

A Task Scheduling Strategy with a Sleep-Delay Timer and a Waking-Up Threshold in Cloud Computing

Shunfu Jin[1][✉], Xiushuang Wang[1], and Wuyi Yue[2]

[1] School of Information Science and Engineering, Yanshan University, Qinhuangdao 066004, People's Republic of China
jsf@ysu.edu.cn, ysuwxs@163.com
[2] Department of Intelligence and Informatics, Konan University, Kobe 658-8501, Japan
yue@konan-u.ac.jp

Abstract. For the purpose of satisfying the response performance of cloud users while reducing the energy consumption in cloud computing, we propose a task scheduling strategy with a sleep-delay timer and a waking-up threshold. According to the stochastic behavior of tasks with the proposed strategy, we establish a synchronous vacation queueing model with vacation-delay and N-policy. Then we derive the average sojourn time of tasks and the energy conservation level of the system in a steady state. Finally, we provide numerical experiments to investigate the impacts of system parameters on performance criteria.

Keywords: Cloud computing · Task scheduling strategy
Sleep-delay timer · Waking-up threshold · Sojourn time · Energy
conservation level

1 Introduction

Cloud computing is a style of computing in which dynamically scalable and virtualized resources are provided as a service over the Internet. However, all servers are responsible for dramatic amounts of energy consumption and carbon dioxide emission, making the creation of a greener cloud environment one of the fundamental challenges in cloud computing. Greener cloud environment aims to reduce the energy utilization and weaken the impact of carbon dioxide emissions on the environment.

Due to its unprecedented computing capability, cloud computing has become a popular paradigm and has attracted attention from a wide range of enterprises. Many scholars have carried out research into the energy management and optimization of cloud computing. Cheng *et al.* designed an algorithm, Minimum Expectation Execution Energy with Performance Constraints (ME^3PC), which can reduce energy consumption effectively while meeting performance

© Springer International Publishing AG, part of Springer Nature 2018
Y. Takahashi et al. (Eds.): QTNA 2018, LNCS 10932, pp. 115–123, 2018.
https://doi.org/10.1007/978-3-319-93736-6_8

constraints [1]. Chen *et al.* proposed a Dynamic Voltage and Frequency Scaling (DVFS) scheme, which can dynamically predict the most suitable voltage and frequency for the multi-core embedded system [2]. Shen *et al.* discussed a genetic algorithm E-PAGA to achieve adaptive regulations for different requirements of energy and performance in cloud tasks [3]. While all these approaches seek to reduce energy consumption, they ignore the issue of VMs staying awake even when no tasks are being processed.

In general, putting idle virtual machines (VMs) in sleep mode during low-load periods is a direct way to reduce power consumption. Kempa investigated a finite-buffer queueing system with vacation, in which the transmission after each sleep period is initialized only after the number of packets in the buffer reaches a threshold [4]. Lawanyashri *et al.* used a vacation and threshold policy to efficiently control the workload level of each VM in the data center. This policy can reduce the energy consumption and cost accordingly [5]. For a greener cloud computing, Singh *et al.* proposed a deep-sleep mode when all the VMs were not being used. With this method, the energy consumption is reduced, while the resource utility is improved [6]. Making idle VMs sleep can reduce energy consumption in a certain degree. However, continually switching VMs can cause response penalties.

Inspired by these observations, we propose a novel energy efficient task scheduling strategy with a sleep-delay timer and a waking-up threshold under the constraint of the quality of experience (QoE) of cloud users. By constructing a two-dimensional continuous-time Markov chain (CTMC), we evaluate the system performance in terms of the average sojourn time of tasks and the energy conservation level of the system.

The remainder of this paper is organized as follows: Sect. 2 describes the task scheduling strategy proposed in this paper and the system model. Section 3 presents the steady-state distribution analysis of the system model. Section 4 gives the performance criteria and provides numerical experiments to evaluate the system by using this proposed strategy. Finally, Sect. 5 draws conclusions of the paper.

2   Task Scheduling Strategy and System Model

In this section, we present the task scheduling strategy that we propose in this paper and present the system model by using this strategy.

2.1   Task Scheduling Strategy

Cloud computing employs distributed hardware and software services. Multiple VMs can be deployed onto one physical machine (PM) through virtualization in cloud computing environments. Groups of VMs distributed across one or more PMs are always awake, even though there are no tasks to be processed. Thus, large amounts of power are wasted.

In cloud data centers, the PM is usually a high configured server. To provide high availability and improve parallel processing capabilities, each VM hosted on one PM works independently under the control of its own operating system. From the perspective of each VM, it is possible to introduce the sleep mode.

In view of this situation, we propose a novel energy efficient task scheduling strategy with a sleep-delay timer and a waking-up threshold for cloud computing. To improve the energy efficiency, we set a critical threshold N. At the completion instant of a sleep period, if the number of tasks waiting in the system buffer reaches or exceeds the threshold, all the VMs will wake up to provide service. Otherwise, another sleep period will be activated and all the VMs will remain asleep. To guarantee the quality of experience (QoE) of cloud users, we set a sleep-delay timer for the PM. The task arriving within the sleep-delay timer will be processed immediately. Only when there are no task arrivals within the sleep-delay timer will all the VMs go to sleep.

In the proposed strategy, there are three operating states for the PM: awake state, sleep state and sleep-delay state.

In the awake state, tasks can be processed continuously following the $first-come\ first-served$ discipline. Once the system becomes empty, a sleep-delay timer will be started, and all the VMs will enter the sleep-delay state. If any new tasks arrive while the sleep-delay timer is on, all the VMs will provide service immediately. Otherwise, all the VMs will go to sleep when the sleep-delay timer expires. Once the PM enters the sleep state, a sleep timer with a random duration will be activated. Tasks arriving while the sleep timer is on will queue in the system buffer. At the completion instant of a sleep period, if the number of tasks queueing in the system buffer is less than the threshold N, another sleep period will be activated and the sleep timer will be restarted. Otherwise, all the VMs will wake up and begin serving all the tasks in the system one by one.

2.2 System Model

By regarding the tasks submitted to the cloud computing system as customers, each VM hosted on the PM as a server, the sleep state as the vacation, the sleep-delay state as the vacation-delay, the waking-up threshold as the N-policy, we establish a synchronous multiple vacation queueing model with a vacation-delay and an N-policy.

Since the behavior of PMs in cloud data centers is stochastically homogeneous, in this paper, we focus on a tagged PM to investigate the task scheduling strategy. The task arrivals are assumed to follow a Poisson process with parameter λ $(0 < \lambda < +\infty)$. Each task requires an independently and identically distributed service time, which also follows an exponential distribution with parameter μ $(0 < \mu + \infty)$. In addition, the time lengths for the sleep-delay timer and the sleep timer are supposed to follow exponential distributions with parameters β $(0 < \beta < +\infty)$ and θ $(0 < \theta < +\infty)$, respectively. The number of VMs in the system is assumed to be k and the system buffer is supposed to be infinite.

The behavior of the system model under consideration can be described in terms of the regular irreducible two-dimensional continuous-time Markov chain

(CTMC) $\{(X_t, Y_t),\ t \geq 0\}$, where $X_t = i\ (i = 0,1,\dots)$ is the number of tasks in the system at the instant t. X_t is also referred to as the system level. $Y_t = j\ (j = 0,1,2)$ is the PM state at the instant t. Y_t is called the PM state. $j = 0$ means the PM is in a sleep state, $j = 1$ means the PM is in an awake state, and $j = 2$ means the PM is in a sleep-delay state.

Let $\pi_{i,j}$ be the steady-state distribution of the two-dimensional CTMC. To satisfy the necessary and sufficient condition for the system model to be stable, we assume the CTMC is positive recurrent. Then we can define $\pi_{i,j}$ as

$$\pi_{i,j} = \lim_{t \to \infty} P\{X_t = i,\ Y_t = j\},\ i = 0,1,\dots,\ j = 0,1,2. \tag{1}$$

We define π_i as the steady-state probability distribution for the system level being equal to i. π_i can be partitioned as follows:

$$\pi_i = (\pi_{i,0}, \pi_{i,1}, \pi_{i,2}),\ i = 0,1,\dots. \tag{2}$$

The steady-state probability distribution $\mathbf{\Pi}$ of the two-dimensional CTMC is composed of $\pi_i\ (i = 0,1,\dots)$. $\mathbf{\Pi}$ is then given as follows:

$$\mathbf{\Pi} = (\pi_0, \pi_1, \dots). \tag{3}$$

3   Model Analysis

In this section, we discuss the state transition of the two-dimensional CTMC and derive the steady-state probability distribution of the system model.

3.1   State Transition

From the model hypothesis mentioned in Subsect. 2.2, we can easily recognize that $\{(X_t, Y_t),\ t \geq 0\}$ is a Markov process. Due to the relationship between the number k of VMs in the system and the value of the threshold N, two forms of the state transition for the CTMC exist. For the case of the $N \leq k$, the state transition of the CTMC is represented in Fig. 1(a). For the case of the $N > k$, the state transition of the CTMC is represented in Fig. 1(b).

From Fig. 1, we can find that the system state transitions occur only between adjacent levels. Hence, the two-dimensional CTMC $\{(X_t, Y_t),\ t \geq 0\}$ can be seen as a type of quasi birth-and-death (QBD) process.

3.2   Steady-State Probability Distribution

For the case in Fig. 1(a), the state transition is repetitive from the level k. For the case in Fig. 1(b), the state transition is repetitive from the level N. The steady-state probability distribution satisfies the matrix geometric solution form as follows:

$$\pi_i = \pi_y R^{i-y},\ i \geq y+1 \tag{4}$$

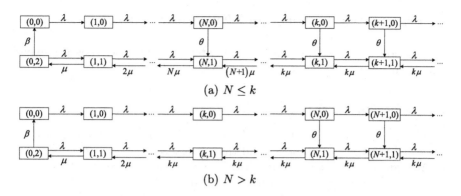

Fig. 1. The state transition of the two-dimensional CTMC presented in this paper.

where y is defined as the maximum of the number k of VMs and the threshold N, i.e., $y = \max\{k, N\}$, $\boldsymbol{R}$ is the rate matrix.

According to the state transition in Subsect. 3.1, the form of the rate matrix $\boldsymbol{R}$ can be assumed as follows:

$$\boldsymbol{R} = \begin{pmatrix} r_{11} & r_{12} & 0 \\ 0 & r_{22} & 0 \\ 0 & 0 & 0 \end{pmatrix} \tag{5}$$

where r_{ij} is the element that need to be solved.

In the matrix geometric solution method shown in [7], the matrix quadratic equation is equivalent to the following set of equations:

$$\begin{cases} k\mu(r_{11}r_{12} + r_{12}r_{22}) + \theta r_{11} - (\lambda + k\mu)r_{12} = 0 \\ k\mu r_{22}^2 - (\lambda + k\mu)r_{22} + \lambda = 0 \\ -(\lambda + \theta)r_{11} + \lambda = 0. \end{cases} \tag{6}$$

By solving Eq. (6), the rate matrix $\boldsymbol{R}$ is derived as follows:

$$\boldsymbol{R} = \begin{pmatrix} \dfrac{\lambda}{\lambda + \theta} & \rho & 0 \\ 0 & \rho & 0 \\ 0 & 0 & 0 \end{pmatrix} \tag{7}$$

where $\rho = \dfrac{\lambda}{k\mu} < 1$.

Applying the normalized condition $(\boldsymbol{\pi}_0, \boldsymbol{\pi}_1, \ldots, \boldsymbol{\pi}_{y-1})\boldsymbol{e} + \boldsymbol{\pi}_y(\boldsymbol{I} - \boldsymbol{R})^{-1}\boldsymbol{e}_1 = 1$ and the Gauss-Seidel method, we can obtain $\boldsymbol{\pi}_i$ $(i = 0, 1, \ldots, y)$. In the normalized condition, $\boldsymbol{e}$ is a $3y \times 1$ vector with ones, and $\boldsymbol{e}_1$ is a 3×1 vector with ones. By substituting $\boldsymbol{\pi}_y$ into Eq. (4), we can obtain $\boldsymbol{\pi}_i$ $(i = y + 1, y + 2, \ldots)$. Then, the steady-state distribution $\boldsymbol{\Pi} = (\boldsymbol{\pi}_0, \boldsymbol{\pi}_1, \ldots)$ of the system can be presented numerically.

4   Performance Criteria and Experimental Results

In this section, we derive the performance criteria and provide numerical experiments to investigate the system performance.

4.1   Performance Criteria

We define the average sojourn time of tasks as the sum of the average waiting time of tasks in the system buffer and the average service time of tasks on the VM. According to the steady-state distribution given in Subsect. 3.2 and using Little's law, the average sojourn time W of tasks can be given as follows:

$$W = \frac{1}{\lambda} \left(\sum_{i=0}^{\infty} i(\pi_{i,0} + \pi_{i,1} + \pi_{i,2}) \right). \tag{8}$$

We then define the energy conservation level of the system as the average energy conservation per unit time with our proposed strategy. Let C_a be the energy consumption per unit time in the awake state or the sleep-delay state. Let C_s be the energy consumption per unit time in the sleep state. Energy can be saved in the sleep state. However, additional energy will be consumed, including the energy consumption C_t for each switching from the sleep state to the awake state, and the energy consumption C_l for each listening. From the perspective of the whole system, the energy conservation level E_v of the system is given as follows:

$$E_v = \sum_{i=0}^{\infty} \pi_{i,0} \times (C_a - C_s) - \sum_{i=1}^{\infty} \pi_{i,0} \times \theta \times C_t - \sum_{i=0}^{\infty} \pi_{i,0} \times \theta \times C_l. \tag{9}$$

4.2   Experimental Results

We provide numerical experiments to investigate the proposed strategy. Referencing to [8], we set the experimental parameters in Table 1.

Table 1. Experimental parameters.

Parameters	Values
Total number n of VMs in the system	20
Service rate μ	0.2 (tasks/ms)
Sleep-delay parameter β	1.2 (times/ms)
Energy consumption level C_s of a sleeping VM	2 (mW)
Energy consumption level C_l of each listening	4 (mW)
Energy consumption level C_a of a busy VM	20 (mW)
Energy consumption level C_t of each switching	12 (mW)

Considering different arrival rate λ of tasks ($\lambda = 0.4$ and $\lambda = 0.7$) as an example, we investigate the change trend of the average sojourn time W of tasks versus the sleep parameter θ for the different threshold N in Fig. 2. For the same threshold, as the sleep parameter increases, the time length of a sleep period gets shorter, the processing of a task is less likely to be delayed. Thus, the average sojourn time of tasks decreases. For the same sleep parameter, as the threshold increases, the time spent waiting in the system buffer becomes longer. Thus, the average sojourn time of tasks increases. Comparing Figs. 2(a) and (b), we find that for the same sleep parameter and the same waking-up threshold, as the arrival rate increases, the number of tasks is more likely to reach the waking-up threshold during the sleep period, thus the sleep period will be terminated sooner. So the average sojourn time of tasks decreases.

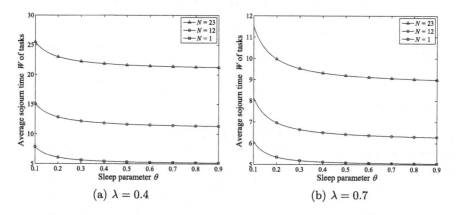

(a) $\lambda = 0.4$ (b) $\lambda = 0.7$

Fig. 2. The average sojourn time of tasks.

Considering different arrival rate λ of tasks ($\lambda = 0.4$ and $\lambda = 0.7$) as an example, we investigate the change trend of the energy conservation level E_v of the system versus the sleep parameter θ for the different threshold N in Fig. 3. For the same threshold, as the sleep parameter increases, the time length of a sleep period gets shorter. Thus the energy conservation within the system decreases. For the same sleep parameter, as the threshold increases, the PM will stay in sleep state longer. Thus the energy conservation within the system increases. Comparing Figs. 3(a) and (b), we find that for the same sleep parameter and the same waking-up threshold, as the arrival rate increases, the PM is less likely to be asleep. Thus, the energy conservation within the system decreases.

In summary, there is a trade-off between the average sojourn time of tasks and the energy conservation level of the system when setting the sleep parameter and the threshold no matter the arrival rate of tasks is larger or smaller. For the applications with higher requirements on response performance, the sleep parameter should be set larger and the waking-up threshold should be set smaller. For the applications with higher requirements on energy conservation level, the

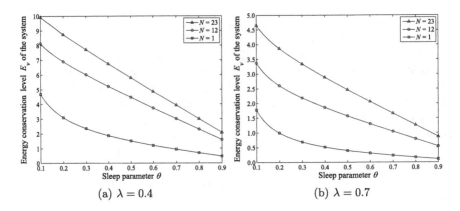

Fig. 3. The energy conservation level of the system.

sleep parameter should be set smaller and the waking-up threshold should be set larger.

5   Conclusions

This paper is original in that it systematically studies the correlation between the response performance and the energy consumption in cloud computing environment. We applied queueing vacation theory as the basic means of modeling the task scheduling strategy in cloud computing. The proposed strategy is representative in realistic cloud scenarios. The presented model can also be extended to evaluate the performance and energy metrics for a more complicated public cloud scenario with numerous different cloud services. The experimental results illustrated changes to the waking-up threshold may potentially lead to energy conservation, and that the use of a sleep-delay timer can bring about task performance enhancement. For further development of our research, we will focus on formulating a cost function to optimize our proposed strategy by considering the tradeoff between the response performance and the energy conservation level.

Acknowledgements. This work was supported in part by National Natural Science Foundation (No. 61472342), Hebei Province Natural Science Foundation (No. F2017203141), China, and was supported in part by MEXT, Japan.

References

1. Cheng, C., Li, J., Wang, Y.: An energy-saving task scheduling strategy based on vacation queueing theory in cloud computing. Tsinghua Sci. Technol. **20**(1), 28–39 (2015)
2. Chen, Y., Chang, M., Liang, W., Lee, C.: Performance and energy efficient dynamic voltage and frequency scaling scheme for multicore embedded system. In: IEEE 6th International Conference on Consumer and Electronics, pp. 58–59 (2016)

3. Shen, Y., Bao, Z., Qin, X., Shen, J.: Adaptive task scheduling strategy in cloud: when energy consumption meets performance guarantee. World Wide Web-Internet Web Inf. Syst. **20**(2), 1–19 (2017)
4. Kempa, W.: Time-dependent analysis of transmission process in a wireless sensor network with energy saving mechanism based on threshold waking up. In: IEEE International Workshop on Signal Processing Advances in Wireless Communications, pp. 26–30 (2015)
5. Lawanyashri, M., Balusamy, B., Subha, S.: Threshold-based workload control for an under-utilized virtual machine in cloud computing. Int. J. Intell. Eng. Syst. **9**(4), 234–241 (2016)
6. Singh, D., Devgan, M.: Multilayer hybrid energy efficient approach in green cloud computing. Int. J. Sci. Res. Dev. **4**(10), 814–818 (2016)
7. Yue, D., Yu, J., Yue, W.: A markovian queue with two heterogeneous servers and multiple vacations. J. Ind. Manag. Optim. **5**(3), 453–465 (2017)
8. Jin, S., Hao, S., Yue, W.: Energy-efficient strategy with a speed switch and a multiple-sleep mode in cloud data centers. In: International Conference on Queueing Theory & Network Applications, pp. 143–154 (2017)

A Clustered Virtual Machine Allocation Strategy Based on an N-Threshold Sleep-Mode in a Cloud Environment

Xiuchen Qie[1], Shunfu Jin[1(✉)], and Wuyi Yue[2]

[1] School of Information Science and Engineering, Yanshan University,
Qinhuangdao 066004, People's Republic of China
qiexiuchen@126.com, jsf@ysu.edu.cn
[2] Department of Intelligence and Informatics, Konan University,
Kobe 658-8501, Japan
yue@konan-u.ac.jp

Abstract. In an effort to improve the energy efficiency of cloud data centers, in this paper, we propose a clustered Virtual Machine (VM) allocation strategy based on an N-threshold sleep-mode in which all the VMs in a cloud data center are clustered into two modules. The VMs in Module I are always awake, whereas the VMs in Module II will go to sleep under a light traffic load. When the number of waiting requests reaches or exceeds the threshold N, sleeping VMs will resume processing requests independently after their corresponding sleep timers expire. Accordingly, we establish an N-policy partially asynchronous multiple vacations queueing model, and derive the energy saving rate of the system. Numerical results are provided to show the efficiency of the proposed strategy in reducing energy consumption.

Keywords: Cloud data center · Clustered VM allocation
N-threshold · Sleep-mode · Energy saving rate

1 Introduction

According to the current "Cisco Global Cloud Index", more than four fifths of the workload in data centers will be handled in cloud data centers by 2019 [1]. As a result, energy efficiency is becoming increasingly important in a cloud environment [2].

The use of a sleep mode improves energy efficiency in cloud data centers [3]. In [4], Duan *et al.* proposed a dynamic idle interval prediction scheme that could estimate the future idle interval length of a CPU and thereby choose the most cost-effective sleep state to minimize the energy consumption during runtime. In [5], Chou *et al.* proposed a fine-grain power management scheme for data center workloads. This scheme dynamically postponed the processing of some requests, created longer idle periods and promoted the use of a deeper sleep mode. In

© Springer International Publishing AG, part of Springer Nature 2018
Y. Takahashi et al. (Eds.): QTNA 2018, LNCS 10932, pp. 124–132, 2018.
https://doi.org/10.1007/978-3-319-93736-6_9

[6], Luo *et al.* proposed a dynamic adaptive scheduling algorithm based on flow preemption and power-aware routing. This algorithm saved energy by decreasing the ratio of low utilization devices and putting more devices into sleep mode. In the literature mentioned above, we note that a sleep mode was only applied to a Physical Machine (PM) rather than a Virtual Machine (VM).

In this paper, taking advantage of virtualization technology in cloud computing, we propose a clustered VM allocation strategy based on an N-threshold sleep-mode, and build an N-policy partially asynchronous multiple vacations queueing model. Then, we evaluate the system performance in terms of the energy saving rate of the system, both mathematically and numerically.

The rest of this paper is organized as follows. In Sect. 2, we propose a clustered VM allocation strategy based on an N-threshold sleep-mode in a cloud environment and build a system model accordingly. In Sect. 3, we analyze the system model using the method of matrix geometric solution. In Sect. 4, we derive the energy saving rate of the system. With numerical experiments, we investigate the system performance with the proposed strategy in Sect. 5. Finally, Sect. 6 concludes the whole paper.

2   VM Allocation Strategy and System Model

In this section, to improve the energy efficiency in a cloud environment, we first propose a clustered VM allocation strategy based on an N-threshold sleep-mode. Then, we develop an N-policy partially asynchronous multiple vacations queueing model to analyze the system with the proposed strategy.

2.1   VM Allocation Strategy

We note that additional energy will be consumed when the VM frequently switches from the sleep state to the awake state, and the system performance will be degraded when all the VMs are put in an imposed sleep-mode. To get around this problem, a clustered VM allocation strategy based on an N-threshold sleep-mode is proposed.

All the VMs in a cloud data center are clustered into two modules, namely, Module I and Module II. The VMs in Module I stay awake and operate on a higher speed. The VMs in Module II will go to sleep independently when there are no requests in the system buffer. At the end of a sleep period, if the requests gathered in the system buffer reaches or exceeds a certain value, namely threshold N, the corresponding VM in Module II will wake up independently and operate on a lower speed. Otherwise, the VM in Module II will restart a sleep timer and begin another sleep period.

In the proposed strategy, all the VMs are dominated by a control server, in which several sleep timers, a request counter, and a VM scheduler are deployed. Each sleep timer determines the time length of a sleep period. The request counter records the number of the requests waiting in the system buffer. Based on the sleep timers and the request counter, the VM scheduler adjusts the system state.

In Fig. 1, we demonstrate the workflow of VMs with the clustered VM allocation strategy based on an N-threshold sleep-mode.

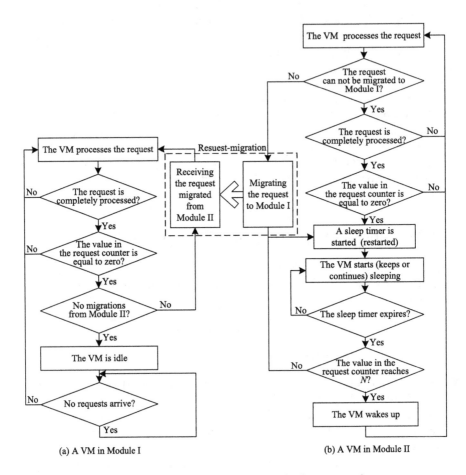

(a) A VM in Module I   (b) A VM in Module II

Fig. 1. The workflow diagram of a VM with the proposed strategy.

2.2 System Model

Regarding a request as a customer, a VM as an independent server, and a sleep period as a vacation, we model the proposed strategy as an N-policy partially asynchronous multiple vacations queueing model.

In this system model, the numbers of the VMs in Module I and Module II are denoted as c and d, respectively. The arrival intervals of requests are assumed to follow an exponential distribution with parameter λ ($\lambda > 0$). The service times of requests processed in Module I and in Module II are assumed to follow exponential distributions with parameters μ_1 ($\mu_1 > 0$) and μ_2 ($0 < \mu_2 < \mu_1$), respectively. Furthermore, the sleep timer length is assumed to follow an exponential distribution with parameter θ ($\theta > 0$). Here, the parameter θ is called the sleeping parameter.

The system model is described with an infinite buffer capacity. Let $S(t) = i, i \in \{0, 1, \ldots\}$ be the total number of requests in the system at instant t. $S(t)$ is also called the system level. Let $J(t) = j, j \in \{0, 1, \ldots, d\}$ be the number of busy VMs in Module II at instant t. $J(t)$ is also called the system stage. Based on the assumptions above, $\{S(t), J(t), t \geq 0\}$ can be regarded as a two-dimensional continuous time Markov chain (CTMC).

We define $\pi_{i,j}$ as the steady-state probability distribution of the system model for the system level being equal to i and the system stage being equal to j.

We define $\boldsymbol{\pi}_i$ as the steady-state probability distribution when the system level is i. The steady-state probability distribution $\boldsymbol{\Pi}$ of the two-dimensional CTMC is composed of $\boldsymbol{\pi}_i$ ($i \geq 0$). $\boldsymbol{\Pi}$ is given as follows:

$$\boldsymbol{\Pi} = (\boldsymbol{\pi}_0, \boldsymbol{\pi}_1, \ldots). \tag{1}$$

3   Model Analysis

Based on the system level, the one-step state transition rate matrix $\boldsymbol{Q}$ of the two-dimensional CTMC $\{(S(t), J(t)), t \geq 0\}$ can be written in a block-tridiagonal form as follows:

$$\boldsymbol{Q} = \begin{pmatrix} A_0 & C_0 & & & & & & & & & \\ B_1 & A_1 & C_1 & & & & & & & & \\ & \ddots & \ddots & \ddots & & & & & & & \\ & & B_c & A_c & C_c & & & & & & \\ & & & B_{c+1} & A_{c+1} & C_{c+1} & & & & & \\ & & & & \ddots & \ddots & \ddots & & & & \\ & & & & & B_{c+d} & A_{c+d} & C & & & \\ & & & & & & B & A_{c+d+1} & C & & \\ & & & & & & & \ddots & \ddots & \ddots & \\ & & & & & & & & B & A_{c+d+N-1} & C \\ & & & & & & & & & B & A & C \\ & & & & & & & & & & \ddots & \ddots & \ddots \end{pmatrix}$$

In the one-step state transition rate matrix $\boldsymbol{Q}$, the sub-matrices $\boldsymbol{B}_k$ are repeated forever starting from the system level $(c + d + 1)$, the sub-matrices $\boldsymbol{A}_k$ are repeated forever starting from the system level $(c + d + N)$, and the sub-matrices $\boldsymbol{C}_k$ are repeated forever starting from the system level $(c + d)$. The repetitive sub-matrices $\boldsymbol{B}_k$, $\boldsymbol{A}_k$ and $\boldsymbol{C}_k$ are represented by $\boldsymbol{B}$, $\boldsymbol{A}$ and $\boldsymbol{C}$, respectively.

B_k $(k = 1, 2, \ldots, c)$, $\boldsymbol{B}_k$ $(k = c+1, c+2, \ldots, c+d)$ and $\boldsymbol{B}$ $(k = c+d+1, c+d+2, \ldots)$ are the one-step state transition rate sub-matrices for the system level k decreasing by one.

$$B_k = k\mu_1, \quad k = 1, 2, \ldots, c,$$

$$\boldsymbol{B}_k = \begin{pmatrix} c\mu_1 & & & & \\ & c\mu_1 + \mu_2 & & & \\ & & \ddots & & \\ & & & c\mu_1 + (x-1)\mu_2 & \\ & & & & c\mu_1 + x\mu_2 \end{pmatrix}, \quad k = c+x, \ x = 1, 2, \ldots, d,$$

$$\boldsymbol{B} = \mathrm{diag}\,(c\mu_1, c\mu_1 + \mu_2, \ldots, c\mu_1 + d\mu_2), \quad k = c+x, \ x = d+1, d+2, \ldots.$$

A_k $(k = 1, 2, \ldots, c)$, $\boldsymbol{A}_k$ $(k = c+1, c+2, \ldots, c+d+N-1)$ and $\boldsymbol{A}$ $(k = c+d+N, c+d+N+1, \ldots)$ are the one step state transition rate sub-matrices for the system level k remaining fixed. For convenience of presentation, we introduce h_y $(h_y = \lambda + c\mu_1 + y\mu_2, \ 0 \leqslant y \leqslant d)$ to simplify the sub-matrices $\boldsymbol{A}_k$ and $\boldsymbol{A}$.

$$A_k = -(\lambda + k\mu_1), \quad k = 0, 1, \ldots, c,$$

$$\boldsymbol{A}_k = \mathrm{diag}\,(-h_0, -h_1, \ldots, -h_x), \quad k = c+x, \ x = 1, 2, \ldots, \min\{N, d\} - 1.$$

For the case of $N > d$,

$$\boldsymbol{A}_k = \mathrm{diag}\,(-h_0, -h_1, \ldots, -h_d), \quad k = c+x, \ x = d, d+1, \ldots, N-1.$$

For the case of $N \leqslant d$,

$$\boldsymbol{A}_k = \begin{pmatrix} -h_0 - d\theta & & & d\theta & & & \\ & \ddots & & & \ddots & & \\ & & -h_{x-N} - (d-(x-N))\theta & (d-(x-N))\theta & & & \\ & & & -h_{x-N} & 0 & & \\ & & & & \ddots & & \ddots \\ & & & & & -h_{x-1} & 0 \\ & & & & & & -h_x \end{pmatrix},$$

$$k = c+x, \ x = N, N+1, \ldots, d-1.$$

$$
\boldsymbol{A}_k =
\begin{pmatrix}
-h_0 - d\theta & & d\theta & & & & \\
& \ddots & & \ddots & & & \\
& & -h_{x-N} - (d-(x-N))\theta & (d-(x-N))\theta & & & \\
& & & -h_{x-N} & 0 & & \\
& & & & \ddots & & \ddots \\
& & & & & -h_{d-1} & 0 \\
& & & & & & -h_d
\end{pmatrix},
$$
$$
k = c + x, x = \max\{N, d\}, \max\{d, N\} + 1, \ldots, d + N - 1.
$$

$$
\boldsymbol{A} =
\begin{pmatrix}
-h_0 - d\theta & d\theta & & & \\
& -h_1 - (d-1)\theta & (d-1)\theta & & \\
& & \ddots & \ddots & \\
& & & -h_{d-1} - \theta & \theta \\
& & & & -h_d
\end{pmatrix},
$$
$$
k = c + x, x = d + N, d + N + 1, \ldots.
$$

C_k $(k = 1, 2, \ldots, c)$, $\boldsymbol{C}_k$ $(k = c+1, c+2, \ldots, c+d-1)$ and $\boldsymbol{C}$ $(k = c+d, c+d+1, \ldots)$ are the one-step state transition rate sub-matrices for the system level k increasing by one.

$$
C_k = \lambda, \ k = 0, 1, \ldots, c,
$$

$$
\boldsymbol{C}_k =
\begin{pmatrix}
\lambda & & & & 0 \\
& \lambda & & & 0 \\
& & \ddots & & \vdots \\
& & & \lambda & 0 \\
& & & & \lambda & 0
\end{pmatrix}, \ k = c + x, \ x = 1, 2, \ldots, d - 1,
$$
$$
\boldsymbol{C} = \mathrm{diag}\,(\lambda, \lambda, \ldots, \lambda), \ k = c + x, \ x = d, d+1, \ldots.
$$

Obviously, the state transitions of the CTMC occur only between adjacent system levels. The two-dimensional CTMC $\{S(t), J(t), t \geq 0\}$ can be seen as a type of Quasi Birth-and-Death (QBD) process.

To analyze the QBD process $\{S(t), J(t), t \geq 0\}$ by using the matrix geometric solution method, we need to solve for the minimal non-negative solution of the matrix quadratic equation $\boldsymbol{R}^2 \boldsymbol{B} + \boldsymbol{R}\boldsymbol{A} + \boldsymbol{C} = \boldsymbol{0}$. This solution is called the rate matrix $\boldsymbol{R}$.

Based on the discussions above, we find that the sub-matrices $\boldsymbol{B}$, $\boldsymbol{A}$ and $\boldsymbol{C}$ are upper-triangular matrices. So, the rate matrix $\boldsymbol{R}$ must be an upper-triangular matrix, and can be explicitly determined.

Applying the Gauss-Seidel method [7], we can obtain π_i $(i = 0, 1, \ldots, c+d+N)$. Based on the matrix geometric solution form $\pi_i = \pi_{c+d+N} R^{i-(c+d+N)}$, $i \geq c+d+N$, we can obtain π_i $(i = c+d+N+1, c+d+N+2, \ldots)$.

4   Performance Measures

We define the energy saving rate of the system as the energy conservation per unit time. Energy saving rate of the system is a measure to compare the total energy consumption in our proposed strategy and that in the conventional strategy. Based on the steady-state probability distribution of the system model given in Sect. 3, the energy saving rate E of the system with our proposed strategy is given as follows:

$$E = E_1 - (E_2 + E_3) \tag{2}$$

where E_1 is the energy saving rate during the sleep period, E_2 and E_3 are the additional energy consumption rates caused by a request-migration and by a listening at the boundary of the sleep period.

$$E_1 = (\omega - \omega_s) \sum_{i=0}^{\infty} \sum_{j=0}^{d} (d-j)\pi_{ij},$$

$$E_2 = \omega_m \sum_{i=c+1}^{c+d} \sum_{j=1}^{d} c\mu_1 \pi_{ij},$$

$$E_3 = \omega_l \sum_{i=0}^{\infty} \sum_{j=0}^{d} \theta(d-j)\pi_{ij}$$

where ω $(\omega > 0)$ is the energy consumption per unit time for a busy VM in Module II. ω_s $(\omega_s > 0)$ is the energy consumption per unit time for a sleeping VM in Module II. ω_m $(\omega_m > 0)$ is the energy consumption for each request-migration. ω_l $(\omega_l > 0)$ is the energy consumption for each listening.

5   Numerical Experiments

In order to quantify the impact of the sleeping parameter on the energy saving rate of the system for the different number of the VMs in Module II and the different thresholds N, we provide numerical experiments. Referencing to [8], we set the experimental parameters as follows: $c + d = 50$, $\lambda = 7.00$ (requests/ms), $\mu_1 = 0.20$ (requests/ms), $\mu_2 = 0.10$ (requests/ms), $\omega = 0.50$ mW, $\omega_s = 0.10$ mW, $\omega_m = 0.50$ mW and $\omega_l = 0.15$ mW.

Figure 2 examines the influence of the sleeping parameter θ on the energy saving rate E of the system for the different number d of the VMs in Module II and the different thresholds N.

From Fig. 2(a), we notice that when the sleeping parameter θ and the threshold N are given, a larger number d of the VMs in Module II will lead to a higher

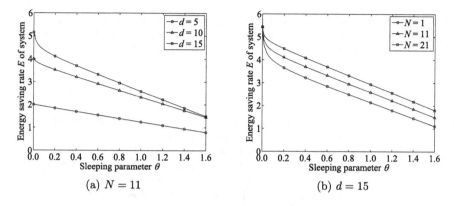

Fig. 2. Energy saving rate E of the system.

energy saving rate E of the system. As the number of the VMs in Module II increases, more VMs have the opportunity to take a sleep, so the energy saving rate of the system improves.

From Fig. 2(b), we notice that when the sleeping parameter θ and the number d of the VMs in Module II are given, a bigger threshold N will lead to a higher energy saving rate E of the system. The higher the threshold N is, the later a VM in Module II will wake up from sleeping, so the VMs in Module II will stay in the sleep state for longer. This results in a higher energy saving rate of the system.

Combining Figs. 2(a) and 2(b), we also observe that for any number d of the VMs in Module II and any value for threshold N, the energy saving rate E of the system decreases as the sleeping parameter θ increases. On the one hand, the larger the sleeping parameter is, the shorter the time length of a sleep period is, and the later a VM in Module II will wake up from sleeping, so less energy will be saved. On the other hand, the larger the sleeping parameter is, the more frequently the VM in Module II listens to the system buffer, so additional energy will be consumed. Therefore, the energy saving rate of the system will decrease.

6   Conclusions

In this paper, we proposed a clustered VM allocation strategy. Considering an N-threshold sleep-mode with the proposed strategy, we established an N-policy partially asynchronous multiple vacations queueing model. The queueing model quantified the effects of the number of VMs in Module II, the threshold N and the sleeping parameter on the energy saving rate of the system. In future research, we aim to extend our study to investigate the average latency of requests and to optimize the proposed strategy by making trade-offs between different performance measures.

Acknowledgements. This work was supported in part by National Natural Science Foundation (No. 61472342), Hebei Province Natural Science Foundation (No. F2017203141), China, and was supported in part by MEXT, Japan.

References

1. Hintemann, R., Clausen, J.: Green Cloud? the current and future development of energy consumption by data centers, networks and end-user devices. In: 4th International Conference on ICT for Sustainability, pp. 109–115 (2016)
2. Jin, X., Zhang, F., Vasilakos, A., Liu, Z.: Green data centers: a survey, perspectives, and future directions (2016). https://arxiv.org/pdf/1608.00687v1.pdf
3. Fan, L., Gu, C., Qiao, L., Wu, W., Huang, H.: GreenSleep: a multi-sleep modes based scheduling of servers for cloud data center. In: International Conference on Big Data Computing and Communications, pp. 368–375 (2017)
4. Duan, L., Zhan, D., Hohnerlein, J.: Optimizing cloud data center energy efficiency via dynamic prediction of CPU idle intervals. In: 8th IEEE International Conference on Cloud Computing, pp. 985–988 (2015)
5. Chou, C., Wong, D., Bhuyan, L.: DynSleep: fine-grained power management for a latency-critical data center application. In: International Symposium on Low Power Electronics and Design, pp. 212–217 (2016)
6. Luo, J., Zhang, S., Yin, L., Guo, Y.: Dynamic flow scheduling for power optimization of data center networks. In: 5th International Conference on Advanced Cloud and Big Data, pp. 57–62 (2017)
7. Jiang, M., Hu, J., Zhao, R., Wei, X., Nie, Z.: Hybrid IE-DDM-MLFMA with Gauss-Seidel iterative technique for scattering from conducting body of translation. Appl. Comput. Electromagn. Soc. J. **30**(2), 148–156 (2015)
8. Jin, S., Ma, X., Yue, W.: Energy-saving strategy for green cognitive radio networks with an LTE-advanced structure. J. Commun. Netw. **18**(4), 610–618 (2016)

Performance Evaluation for a Registration Service with an Energy Efficient Cloud Architecture

Haixing Wu[1], Shunfu Jin[1(✉)], Wuyi Yue[2], and Yutaka Takahashi[3]

[1] School of Information Science and Engineering, Yanshan University,
Qinhuangdao 066004, People's Republic of China
wuhaixing1112@163.com, jsf@ysu.edu.cn
[2] Department of Intelligence and Informatics,
Konan University, Kobe 658-8501, Japan
yue@konan-u.ac.jp
[3] Graduate School of Informatics, Kyoto University, Kyoto 606-8225, Japan
takahashi@i.kyoto-u.ac.jp

Abstract. Cloud computing allows application providers to seamlessly scale services and enables users to adaptively scale usage. Cloud vendors always provide a free service to appeal to more anonymous users. In this paper, we propose a sleep-mode based cloud architecture, in which a free service and an optional registration service are provided on the same server. Regarding the free service as the first service, the registration service as the second optional service and the sleep state as the vacation, we establish an asynchronous multiple-vacation queueing model with a second optional service. We construct a three-dimensional Markov chain to derive the steady-state distribution of the queueing model, and estimate the average response time of anonymous users and the energy saving rate of system. Finally, we provide numerical results to investigate the trade-off between difference performance measures.

Keywords: Cloud computing · Registration service
Second optional service queue · Energy saving rate
Average response time

1 Introduction

Cloud computing is offering utility-oriented Information Technology (IT) services to users worldwide [1]. In order to appeal to more users, cloud vendors always provide free service to anonymous users. If an anonymous user is satisfied with the free service and believes they are likely to receive better service next time, the anonymous user may well register as a VIP (Very Important Person) user. On the other hand, in cloud computing systems, the energy consumption of the under-utilized resources accounts for a substantial amount of the actual energy use [2]. Therefore, how to provide a registration service along with energy efficient cloud architecture is an important issue for cloud vendors.

© Springer International Publishing AG, part of Springer Nature 2018
Y. Takahashi et al. (Eds.): QTNA 2018, LNCS 10932, pp. 133–141, 2018.
https://doi.org/10.1007/978-3-319-93736-6_10

A scientific evaluation of system performance helps to run a cloud system well. Queueing theory with a second optional service is suitable for modeling the registration service in cloud systems. Queueing theory with a vacation mechanism is suitable for modeling the sleep mode in cloud systems. Queueing theory with second optional service was first formulated by Madan [3]. Following Madan, there were various papers on queueing models with a second optional service. Wei et al. discussed a discrete-time Geom/G/1 retrial queue with balking customers and a second optional service, where the retrial time followed a geometrical distribution [4]. The vacation model terminology first appeared in the 1970s. Doshi wrote an excellent survey paper on vacation models [5]. Numerous papers on vacation models have appeared since that time. Jain et al. considered an asynchronous vacation policy for a multi-server repair problem with a server breakdown and two types of spares [6]. All the aforementioned papers do not study a queueing model combining the second optional service and the vacation together.

In this paper, we firstly propose a sleep-mode based cloud architecture where a free service and a registration service are provided on the same server. The newly vacated server will enter the sleep state once there are no users waiting in the system buffer. We build an asynchronous multiple-vacation queueing model with a second optional service to investigate the system performance of the proposed cloud architecture. Then, we construct a three-dimensional Markov chain from the perspective of the total number of anonymous users, the number of servers running normally and the number of anonymous users applying for the registration service to analyze the queueing model. Moreover, we evaluate the system performance in terms of the average response time of anonymous users and the energy saving rate of the system in a cloud environment.

The rest of this paper is organized as follows: In Sect. 2, we propose a sleep-mode based cloud architecture with a second optional service, and establish a queueing model accordingly. In Sect. 3, we analyze the queueing model using the method of matrix-geometric solution. In Sect. 4, we analyze the average response time of anonymous users and the energy saving rate of system. In Sect. 5, we demonstrate the influence of system parameters on the system performance with numerical results. Finally, we summarize the conclusions in Sect. 6.

2   System Model

In this section, we propose a sleep-mode based cloud architecture with a registration service. Then, we establish an asynchronous multiple-vacation queueing model with a second optional service.

2.1   Cloud Architecture

It is a common practice for cloud vendors to offer a free service to attract new anonymous users. In conventional cloud computing systems, all the virtual machines (VMs) always stay awake even if there are no users to be serviced. This results in a large amount of wasted energy.

Considering the energy efficiency, we present a sleep-mode based cloud architecture with tasks initiated by ordinary cloud users. In cloud environment, the configuration of physical machines (PMs) is usually very high. Several VMs are deployed to a PM and each VM runs its own operating system independently. This makes it possible to implement sleep mode at the VM level. Anonymous users receive an essential free service and an optional registration service from one of the VMs.

(1) When an anonymous user enters the system, the anonymous user will queue in the system buffer waiting for the free cloud service. Once there is at least one newly vacated VM or one newly woken VM on any PMs, this VM will be allocated to the first anonymous user queueing in the system buffer by the task scheduler. Then the anonymous user allocated the VM will receive the free cloud service.

(2) After the completion of the free cloud service, the anonymous user selects whether to receive the registration service according to the service satisfaction. If the anonymous user opts to register as a VIP user after experiencing a registration process, the user has to pay a reasonable fee and will get a better service next time. Otherwise, the anonymous use will leave the system directly and remain an anonymous user.

(3) If there are no anonymous users waiting in the system buffer and a user departs from a VM, i.e., a VM is vacated, the VM will enter the sleep mode. Once the VM enters the sleep state, a sleep timer with a random durations will be activated to control the time length of the sleep period. At the end of the sleep period, if there are no anonymous users in the system buffer, another sleep timer will be activated, and the VM will begin another sleep period. Otherwise, the VM will return to the active state and wake up for serving the anonymous users in the system buffer.

It is necessary to mathematically evaluate the cloud service with the proposed architecture.

2.2   System Model

Regarding the free service as the first essential service, the registration service as the second optional service, the sleep mode as the vacation and a VM as a server, we establish an asynchronous multiple-vacation queueing model with a second optional service.

The buffer in the system is supposed to be infinite. The total number of VMs in the system is c. Let the random variable $N(t) = i$, $i \in \{0, 1, \ldots\}$ be the total number of anonymous users in system at instant t. Let the random variable $Y(t) = j$, $j \in \{0, 1, \ldots, \min(i, c)\}$ be the number of servers running normally at instant t. Let the random variable $S(t) = k$, $k \in \{0, 1, \ldots, \min(j, c)\}$ be the number of anonymous users who are experiencing the registration service at instant t. $N(t)$ is called the system level, $Y(t)$ is called the system state and

$S(t)$ is called the system phase. $\{N(t), Y(t), S(t), \ t \geq 0\}$ constitutes a three-dimensional continuous-time stochastic process with state space $\boldsymbol{\Omega}$ as follows:

$$\boldsymbol{\Omega} = \{(i, j, k) \mid i \in \{0, 1, \ldots\}, \ j \in \{0, 1, \ldots, \min(i, c)\}, \\ k \in \{0, 1, \ldots, \min(j, c)\}\}. \tag{1}$$

In this model, we assume that the arrival intervals of anonymous users follow an exponential distribution with parameter λ ($\lambda > 0$). We assume that the free service time and the registration service time of an anonymous user follow exponential distributions with mean service times $1/\mu_1$ ($\mu_1 > 0$) and $1/\mu_2$ ($\mu_2 > 0$), respectively. Moreover, we assume that an anonymous user selects the registration service with probability q or opts not to select the registration service with probability $\bar{q}$ ($\bar{q} = 1 - q$). Furthermore, we assume that the time length of the sleep timer follows an exponential distribution with parameter θ ($\theta > 0$). The traffic load ρ of the system model can be given as follows:

$$\rho = \frac{\lambda(\mu_2 + q\mu_1)}{c\mu_1\mu_2}. \tag{2}$$

Based on the assumptions above, we conclude that the stochastic process $\{N(t), Y(t), S(t), \ t \geq 0\}$ is a three-dimensional continuous-time Markov chain (CTMC).

We define π_{ijk} as the steady-state probability distribution of the three-dimensional CTMC for the system level being equal to i, the system state being equal to j, and the system phase being equal to k. π_{ijk} is then given as follows:

$$\pi_{ijk} = \lim_{t \to \infty} P\{N(t) = i, Y(t) = j, S(t) = k\}, \ (i, j, k) \in \boldsymbol{\Omega}. \tag{3}$$

We define the vector $\boldsymbol{\pi}_i$ as the steady-state probability distribution for the system level being equal to i. The steady-state probability distribution $\boldsymbol{\Pi}$ of the three-dimensional CTMC is composed of $\boldsymbol{\pi}_i$ ($i = 0, 1, \ldots$). $\boldsymbol{\Pi}$ is given as follows:

$$\boldsymbol{\Pi} = (\boldsymbol{\pi}_0, \boldsymbol{\pi}_1, \ldots). \tag{4}$$

3   Model Analysis

In this section, we first investigate the transition rate matrix of the three-dimensional CTMC. Then, we derive the steady-state distribution of the system model.

3.1   Transition Rate Matrix

The necessary step in analyzing the steady-state distribution of the system model is to construct the transition rate matrix.

Let $\boldsymbol{Q}$ be the one-step state transition rate matrix of the three-dimensional CTMC $\{N(t), Y(t), S(t), \ t \geq 0\}$. Let $\boldsymbol{Q}_{x,y}$ be the one-step state transition rate sub-matrix for the system level changing from x ($x = 0, 1, \ldots$) to y ($y = 0, 1, \ldots$). For convenience of presentation, we denote $\boldsymbol{Q}_{x,x-1}$ as $\boldsymbol{B}_x$, $\boldsymbol{Q}_{x,x}$ as $\boldsymbol{A}_x$, and $\boldsymbol{Q}_{x,x+1}$ as $\boldsymbol{C}_x$. We discuss $\boldsymbol{B}_x$, $\boldsymbol{A}_x$, and $\boldsymbol{C}_x$ in respect to three cases.

(1) System level decreases:

$1 \leq x \leq c$ means that the number of anonymous users is no more than the number of VMs.

$x > c$ means that the number of anonymous users is more than the number of VMs in the system.

(2) System level remains fixed:

$x = 0$ means that there are no anonymous users in the system and all the VMs are in the sleep state.

$x \geq 1$ means that there is at least one anonymous user in the system.

(3) System level increases:

No matter how many anonymous users are there in the system, how many VMs that are working normally, or how many anonymous users who are applying for the registration service, as long as there is one anonymous user arriving at the system, the system level increases by one, while the system state and the system phase remain fixed, and the transition rate is λ.

For this, we write Q as follows:

$$Q = \begin{pmatrix} A_0 & C_0 & & & & & & \\ B_1 & A_1 & C_1 & & & & & \\ & B_2 & A_2 & C_2 & & & & \\ & & \ddots & \ddots & \ddots & & & \\ & & & B_{c-1} & A_{c-1} & C_{c-1} & & \\ & & & & B_c & A & C & \\ & & & & & B & A & C \\ & & & & & & \ddots & \ddots & \ddots \end{pmatrix}.$$

The block-tridiagonal structure of Q shows that the state transitions occur only between adjacent levels. Hence, the three-dimensional CTMC $\{N(t), Y(t), S(t),\ t \geq 0\}$ can be seen as a type of Quasi Birth-and-Death (QBD) process.

3.2   Steady-State Distribution

To analyze this QBD process, we need to solve the matrix quadratic equation $R^2 B + RA + C = 0$ for the minimal non-negative solution R, and the spectral radius $SP(R) < 1$.

Using the rate matrix R obtained from the matrix quadratic equation, we construct a square matrix $B[R]$ as follows:

$$B[R] = \begin{pmatrix} A_0 & C_0 & & & & \\ B_1 & A_1 & C_1 & & & \\ & B_2 & A_2 & C_2 & & \\ & & \ddots & \ddots & \ddots & \\ & & & B_{c-1} & A_{c-1} & C_{c-1} \\ & & & & B_c & RB + A \end{pmatrix}. \tag{5}$$

Using the method of matrix-geometric solution [7] and normalized condition, we can give a set of equations as follows:

$$\begin{cases} (\boldsymbol{\pi}_0, \boldsymbol{\pi}_1, \ldots, \boldsymbol{\pi}_c) B[\boldsymbol{R}] = 0 \\ (\boldsymbol{\pi}_0, \boldsymbol{\pi}_1, \ldots, \boldsymbol{\pi}_{c-1}) e + \boldsymbol{\pi}_c (\boldsymbol{I} - \boldsymbol{R})^{-1} e_1 = 1 \end{cases} \tag{6}$$

where e is a $\left(\frac{c(c+1)(c+2)}{2}\right) \times 1$ vector with ones, and e_1 is a $\left(\frac{(c+1)(c+2)}{2}\right) \times 1$ vector with ones.

By using the Gauss-Seidel method to solve Eq. (6), we can obtain $\boldsymbol{\pi}_0, \boldsymbol{\pi}_1, \boldsymbol{\pi}_2,$ $\ldots, \boldsymbol{\pi}_c$. From the structure of the one step state transition rate matrix $\boldsymbol{Q}$, we know that $\boldsymbol{\pi}_i$ $(i = c+1, c+2, \ldots)$ satisfies the matrix-geometric solution form as follows:

$$\boldsymbol{\pi}_i = \boldsymbol{\pi}_c \boldsymbol{R}^{i-c}, \ i \ge c. \tag{7}$$

By substituting $\boldsymbol{\pi}_c$ obtained from Eq. (6) into (7), we can obtain $\boldsymbol{\pi}_i$ $(i = c+1, c+2, \ldots)$. Then, the steady-state distribution $\boldsymbol{\Pi} = (\boldsymbol{\pi}_0, \boldsymbol{\pi}_1, \ldots)$ of the system can be given numerically.

4   Performance Measures

We define the response time of a user as the duration from the instant this user arrives at the system to the instant this user completes service and departs from the system. We note that the response time of a user includes the time period waiting in the system buffer and the time period getting service from the system. Following Little's law, the average waiting time $E[W]$ of anonymous users is then given as follows:

$$\begin{aligned} E[W] &= \frac{1}{\lambda} E[L] \\ &= \frac{1}{\lambda} \left(\sum_{i=c}^{\infty} \sum_{j=0}^{c} \sum_{k=0}^{j} (i - c) \pi_{ijk} \right). \end{aligned} \tag{8}$$

For an anonymous user who selects the registration service, the average service time $E[X]$ is the sum of the average free service time and the average registration service time, i.e., $E[X] = \frac{1}{\mu_1} + \frac{1}{\mu_2}$.

The average response time $E[T]$ of an anonymous user who selects the registration service is then given as follows:

$$\begin{aligned} E[T] &= E[W] + E[X] \\ &= \frac{1}{\lambda} \left(\sum_{i=c}^{\infty} \sum_{j=0}^{c} \sum_{k=0}^{j} (i - c) \pi_{ijk} \right) + \frac{1}{\mu_1} + \frac{1}{\mu_2}. \end{aligned} \tag{9}$$

We define the energy saving rate as the energy conservation per unit time for a system with the proposed sleep-mode based cloud architecture. During the active state of the VM, the energy will be consumed normally, while during the sleep state of the VM, the energy will be saved. Moreover, additional energy will be consumed when a VM switches from the sleep state to the active state.

Let V_{a1} be the energy consumption per unit time of a VM providing the free service, V_{a2} be the energy consumption per unit time of a VM providing the registration service, V_s be the energy consumption per unit time of a VM in the sleep state, and V_t be the energy consumption for each VM switching from the sleep state to the active state. The energy saving rate $E[S]$ of the system is

$$E[S] = (V_a - V_s) \sum_{i=0}^{\infty} \sum_{j=0}^{c} \sum_{k=0}^{j} (c - j)\pi_{ijk} - V_t \sum_{i=0}^{\infty} \sum_{j=0}^{c} \sum_{k=0}^{j} (c - j)\pi_{ijk} \times \theta \quad (10)$$

where $V_a = qV_{a1} + (1 - q)(V_{a1} + V_{a2})$.

5   Numerical Experiments

In order to evaluate the system performance of the proposed sleep-based cloud architecture, we provide numerical experiments with analysis.

Generally speaking, for a massive cloud data center, there are many VMs in a PM. The service ability of a VM is stronger, however, the energy consumption is greater. By setting $c = 10$, $\mu_1 = 0.5$, $\mu_2 = 0.2$, $q = 0.5$, $V_{a1} = 5$ mW, $V_{a2} = 4$ mW, $V_s = 0.5$ mW and $V_t = 6$ mW as an example, we show how the average response time $E[T]$ of anonymous users who select the registration service and the energy saving rate of the system changes versus the arrival rate λ of the anonymous users for different the sleep parameter θ in Figs. 1 and 2, respectively.

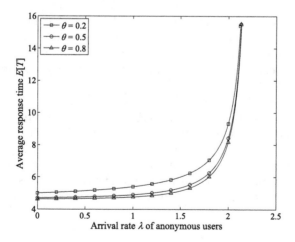

Fig. 1. Change trend of the average response time of anonymous users who select the registration service.

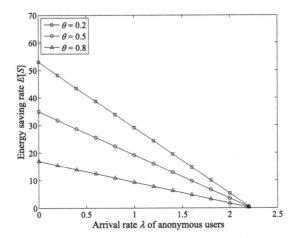

Fig. 2. Change trend of the energy saving rate.

As can be seen from Fig. 1, for the same sleep parameter θ, as the arrival rate λ of anonymous users increases, the waiting time of the anonymous users including the anonymous users who will select the registration service later in the system buffer will be longer. Therefore, the average response time $E[T]$ of the anonymous users who select the registration service increases. For the same arrival rate λ of the anonymous users, as the sleep parameter θ increases, the time length of a sleep period gets shorter, the anonymous users arriving during the sleep period can be served earlier. Therefore, the average response time $E[T]$ of the anonymous users who select the registration service will decreases.

From Fig. 2 we find that with the same sleep parameter θ, as the arrival rate λ of the anonymous users increases, the VMs are less likely to be asleep. Therefore, the energy saving rate $E[S]$ of the system decreases. For the same arrival rate λ of the anonymous users, as the sleep parameter θ increases, the time length of a sleep period will become shorter. Therefore, the energy saving rate $E[S]$ of the system decreases.

For a larger sleep parameter, the average response time of the anonymous users selected the registration service is shorter, but the energy saving rate of the system is lower. For a smaller sleep parameter, the energy saving rate of the system is higher, but the average response time of the anonymous users selected the registration service is longer. We understood that there is a trade-off between the average response time of the anonymous users who select the registration service and the energy saving rate of the system when setting the sleep parameter.

6   Conclusions

In this paper, we proposed a sleep-mode based cloud architecture. Accordingly, we presented a method to model and evaluate the proposed cloud architecture by establishing an asynchronous multiple-vacation queueing model with a second optional service. We provided numerical results to investigate the impact of the sleep parameter on the average response time of the anonymous users who select the registration service and the energy saving rate of the system. In future work, this research would be extended to the optimization for the sleep parameter.

Acknowledgements. This work was supported in part by Hebei Province Natural Science Foundation (No. F2017203141), China, and was supported in part by MEXT and JSPS KAKENHI Grant (Nos. JP17H01825 and JP26280113), Japan.

References

1. Hussein, S., Alkabani, Y., Mohamed, H.: Green cloud computing: datacenters power management policies and algorithms. In: Proceedings of the International Conference on Computer Engineering and Systems, pp. 421–426 (2015)
2. Dhanwate, R., Bhagat, V.: Improving energy efficiency on android using cloud based services. Int. J. Adv. Res. Comput. Sci. Manag. Stud. **3**(5), 75–79 (2015)
3. Madan, K.: An M/G/1 queue with second optional service. Queueing Syst. **34**(1), 37–46 (2000)
4. Wei, C., Cai, L., Wang, J.: A discrete-time Geom/G/1 retrial queue with balking customers and second optional service. OPSEARCH **53**(2), 344–357 (2016)
5. Doshi, B.: Queueing systems with vacationsa survey. Queueing Syst. **1**(1), 29–66 (1986)
6. Jain, A., Jain, M.: Multi server machine repair problem with unreliable server and two types of spares under asynchronous vacation policy. Int. J. Math. Oper. Res. **10**(3), 286–315 (2017)
7. Jin, S., Hao, S., Yue, W.: Energy-efficient strategy with a speed switch and a multiple-sleep mode in cloud data centers. In: Yue, W., Li, Q.L., Jin, S., Ma, Z. (eds.) QTNA 2017. LNCS, vol. 10591, pp. 143–154. Springer, Cham (2017). https://doi.org/10.1007/978-3-319-68520-5_9

Queueing Models for Wireless Communication

Spatial Modelling and Analysis of WLAN with Poisson Point Process

Youngrock Oh and Ganguk Hwang$^{(\boxtimes)}$

Department of Mathematical Sciences, Korea Advanced Institute of Science and
Technology, Daejeon, Republic of Korea
jack90@kaist.ac.kr, guhwang@kaist.edu

Abstract. Wireless Local Area Networks (WLAN) has been in the spotlight as a potential solution to solve the exponentially growing demand of wireless services due to its wide availability and there have been a plenty of research works on WLAN including mathematical modeling and analysis. However, the spatial modeling and analysis of WLAN have received little attention due to the complexity in the dynamics of WLAN. In this work we tackle this issue and provide a spatial modeling and analysis based on Poisson point process to investigate the impact of the spatial distribution on the performance of WLAN. Through our spatial modeling and analysis we verify the conditions where the independence assumption on successful transmissions of data packets holds, which is most widely assumed in most of the previous works on WLAN.

Keywords: Poisson Point Process · Spatial modeling
Transmission success probability · WLAN

1   Introduction

Currently, the era of the Internet of Things (IoT) has opened and tons of devices are connected over wireless links. As one of the most promising solutions to meet the increasing wireless communication demand, the IEEE 802.11 Wireless Local Area Network (WLAN) has been in the spotlight for decades due to its easy implementation and hence wide availability. Since there are a plenty of wireless devices over a wide range of the network spectrum, multiple users in WLAN are forced to coexist in the same channel. So efficient channel sharing in WLAN has been one of important research issues and a number of studies have attempted to analyze and improve the performance of WLAN [1–3].

In the performance modeling and analysis of WLAN, if we take into consideration the dynamics of real networks in the model such as randomness in the number of nodes and the locations of nodes, the complexity of the analysis is substantially increased. This is one of the main reasons why most of the previous works on the performance modeling and analysis of WLAN assume fixed number of nodes, e.g., [1–3], and do not consider the locations of nodes in WLAN that

© Springer International Publishing AG, part of Springer Nature 2018
Y. Takahashi et al. (Eds.): QTNA 2018, LNCS 10932, pp. 145–159, 2018.
https://doi.org/10.1007/978-3-319-93736-6_11

obviously and significantly affect the packet transmissions through channel fading and interference, e.g., [4–7]. However, if we do not consider in performance modeling the main characteristics in WLAN such as the randomness in the number and locations of nodes, the resulting investigation from the model might not be true in a practical WLAN. This is the motivation of this work where we use the theory of stochastic geometry to consider the randomness in the number and locations of nodes.

Stochastic geometry is a good and tractable mathematical tool that is based on point process theory that well captures the randomness mentioned above in WLAN by a spatial distribution of nodes. For this reason, stochastic geometry is widely used to analyze the performance of wireless networks [8–13]. Regarding the study of WLAN based on stochastic geometry [12,13], [12] investigates the performance of mobile data offloading through Wi-Fi where the deployment of Wi-Fi is modeled as a Poisson Point Process (PPP). The work in [13] analyzes the performance of downlink traffic in WLAN by modeling the locations of APs and nodes with a PPP.

In most of previous works on the modeling and analysis of WLAN, it is widely assumed that individual transmission results are independent [1–7,14–16] because the performance results based on the assumption are well matched with simulation results and the assumption makes the analysis eminently simple. However, the simulation in the previous works usually consider a fixed number of nodes or do not consider the locations of nodes and channel fading. So it is desirable to check if the performance results based on the assumption still remains valid when we consider the spatial distribution of nodes in the mathematical and simulation models. More specifically, we focus on the time interval between two successive transmissions of an arbitrary transmitter to investigate whether the spatial distribution of nodes affects the validity of the independence assumption. It is reasonable to assume that the locations of nodes remain *unchanged* during the time interval because one packet transmission time is relatively very short compared to the moving speed of a user. This naturally raises a question that the results of the two consecutive transmissions may be dependent.

To tackle this issue, we develop a mathematical model of WLAN based on stochastic geometry where the spatial distribution of nodes is captured by a PPP. For the channel access in WLAN, we consider the Renewal Access Protocol (RAP) which is recently proposed to improve the throughput and fairness performance of the IEEE 802.11 Distributed Coordination Function (DCF) [14]. We analyze the performance of WLAN based on the model and investigate the impact of the randomness in practical WLANs.

Our main contributions in this work are summarized as follows.

- We develop a mathematical model where the spatial distribution of nodes in WLAN is captured. We also consider in our mathematical model the contention window size for selecting a backoff counter that is ignored in the model in [13].
- Using our mathematical model, we derive analytical expressions on the probability of having a successful packet transmission and the probability of having

consecutive successful packet transmissions. With our analytical results, we investigate the impact of the spatial distribution of nodes on the probability of having a successful packet transmission.

- It is widely assumed in the performance analysis of WLAN that individual packet transmissions result in either success or failure independently, which makes the analysis simple and tractable. Comparing the ratio of two probabilities mentioned above, we investigate whether the independence assumption is valid. What we find is that the independent assumption is only valid when the network environment is relatively good, i.e., the expected number of nodes in the network is relatively small.

The rest of the paper is organized as follows. In Sect. 2 we provide our system modeling. In Sect. 3 we derive the probability of having a successful packet transmission and the probability of having consecutive successful packet transmissions. In Sect. 4 we provide simulation studies to validate our analysis and investigate the performance behaviors of WLAN. Finally, our conclusions are provided in Sect. 5.

2   System Modeling

We consider a wireless network consisting of homogeneous nodes equipped with the RAP proposed in [14]. The RAP is the same as the IEEE 802.11 DCF except the backoff stage and it performs as follows. When a node in the network has a packet to transmit, it selects its own backoff counter value according to a priori given selection distribution on $[1, W]$ where W denotes the window size. The backoff counter value of each node is decremented by 1 if there are no packet transmissions in the network during a fixed time length or is frozen until the end of packet transmissions if a packet transmission occurs in the network. The node transmits its packet whenever its backoff counter value becomes 0. After a packet transmission, the node selects a new backoff counter value with the selection distribution.

We consider an arbitrary node and its embedded time epochs where the backoff counter value of the node is decremented by 1, is frozen (due to the transmissions by some other nodes), or is renewed by a new value. A time interval between two consecutive embedded epochs is called a slot. We assume the perfect sensing, i.e., each node has correct information whether the channel is idle or not. We remove all busy slots (where there occur packet transmissions in the network) as in [14]. The remaining idle slots are indexed by $t = 0, 1, 2, \cdots$. For simplicity and by abuse of notation we simply use the term *slots* to denote the remaining idle slots. When the probability mass function of the selection distribution is denoted by $\{p_1, p_2, \cdots, p_W\}$, the transition probability matrix for the backoff counter of a node between two consecutive slots, denoted by $\mathbf{P}$, is given by

$$\mathbf{P} = \begin{array}{c} \\ 1 \\ 2 \\ 3 \\ \vdots \\ W \end{array} \begin{array}{cccccc} 1 & 2 & 3 & \cdots & W-1 & W \\ \begin{pmatrix} p_1 & p_2 & p_3 & \cdots & p_{W-1} & p_W \\ 1 & 0 & 0 & \cdots & 0 & 0 \\ 0 & 1 & 0 & \cdots & 0 & 0 \\ \vdots & \vdots & \vdots & \ddots & \vdots & \vdots \\ 0 & 0 & 0 & \cdots & 1 & 0 \end{pmatrix} \end{array}.$$

For analysis we consider two nodes in the network where one node, called the tagged transmitter, is to transmit and the other node, called the tagged receiver, is to receive the transmitted packet from the tagged transmitter. We assume that the tagged receiver is located at the origin and the tagged transmitter is located at $\mathbf{z}$ in $\mathbb{R}^2$. All the other transmitters are distributed according to a Homogeneous Poisson Point Process (HPPP) with intensity λ in $\mathbb{R}^2$.

Let $g(r)$ be the path loss function for $r \geq 0$ where r is the distance between two nodes involving the transmission of interest. We assume that the path loss function $g(r)$ is given by $g(r) = r^{-\alpha}$ where $\alpha > 2$ is the path loss exponent determined by the environment (outdoor, indoor, urban, suburb, etc.). Let $h(k)$ denote the fading power received at the tagged receiver from the tagged transmitter in slot k. Similarly, let $h_{\mathbf{x}}(k)$ denote the fading power received at the tagged receiver from a transmitter located at $\mathbf{x}$ in slot k. We assume that all fading powers are independent and exponentially distributed with mean $\frac{1}{\mu}$ [12,13]. Let $N(k)$ and $I(k)$ denote the noise power and the interference at the tagged receiver in slot k. We assume that the noise power process $\{N(k), k \geq 0\}$ forms independent and identically distributed random variables.

Let ρ be the Signal-to-Interference-plus-Noise Ratio (SINR) threshold value to have a successful transmission at the tagged receiver. We assume that there occurs a transmission from the tagged transmitter to the tagged receiver just before the end of slot $t = 0$. Then the tagged transmitter selects a new backoff counter value according to a given selection distribution. If the new backoff counter value is k, then the next transmission from the tagged transmitter occurs just before the end of the k-th slot (for simplicity we say "in slot k" from now on) by our modeling assumption. To derive the joint probability of having consecutive successful packet transmissions, we first derive, for $k \geq 1$

$$P\left\{ \frac{h(0)g(||\mathbf{z}||)}{N(0) + I(0)} > \rho, \frac{h(k)g(||\mathbf{z}||)}{N(k) + I(k)} > \rho, BC = k \right\}$$

where BC denotes the new backoff counter value selected by the tagged transmitter after a transmission in slot 0 and $||\mathbf{v}||$ denotes the length of a vector $\mathbf{v}$ in $\mathbb{R}^2$.

$$P\left\{\frac{h(0)g(\|\mathbf{z}\|)}{N(0)+I(0)}>\rho,\frac{h(k)g(\|\mathbf{z}\|)}{N(k)+I(k)}>\rho, BC=k\right\}$$

$$=P\{BC=k\}P\left\{h(0)>\frac{\rho}{g(\|\mathbf{z}\|)}(N(0)+I(0)), h(k)>\frac{\rho}{g(\|\mathbf{z}\|)}(N(k)+I(k))\right\}$$

$$=P\{BC=k\}E\left[P\left\{h(0)>\frac{\rho}{g(\|\mathbf{z}\|)}(N(0)+I(0)),\right.\right.$$

$$\left.\left.h(k)>\frac{\rho}{g(\|\mathbf{z}\|)}(N(k)+I(k))\right|\Phi, N(0), N(k), I(0), I(k)\right\}\right]$$

$$=P\{BC=k\}E\left[e^{-\mu\rho(\|\mathbf{z}\|)(N(0)+I(0))}e^{-\mu\rho(\|\mathbf{z}\|)(N(k)+I(k))}\right]$$

$$=P\{BC=k\}E\left[e^{-\mu\rho(\|\mathbf{z}\|)N(0)}\right]E\left[e^{-\mu\rho(\|\mathbf{z}\|)N(k)}\right]E\left[e^{-\mu\rho(\|\mathbf{z}\|)(I(0)+I(k))}\right]$$

$$(1)$$

where $\rho(r):=\frac{\rho}{g(r)}$.

3   Performance Analysis

In this section, we derive the probability of having a successful packet transmission and the probability of having consecutive successful packet transmissions. To this end, we first compute the last term of (1) as follows. From the Laplace functional of the independently marked PPP [17] we have

$$E\left[e^{-\mu\rho(\|\mathbf{z}\|)(I(0)+I(k))}\right]$$

$$=E\left[e^{-\mu\rho(\|\mathbf{z}\|)\sum_{\mathbf{x}\in\Phi}(1_{\{\mathbf{X}\in\Phi(0)\}}h_{\mathbf{x}}(0)+1_{\{\mathbf{x}\in\Phi(k)\}}h_{\mathbf{x}}(k))g(\|\mathbf{X}\|)}\right]$$

$$=\exp\left(-\int_{\mathbb{R}^2}1-E\left[e^{-\mu\rho(\|\mathbf{z}\|)(1_{\{\mathbf{x}\in\Phi(0)\}}h_{\mathbf{x}}(0)+1_{\{\mathbf{x}\in\Phi(k)\}}h_{\mathbf{x}}(k))g(\|\mathbf{x}\|)}\right]\Lambda(d\mathbf{x})\right).$$

$$(2)$$

where $1_{\{.\}}$ denotes the indicator function, Λ is intensity measure with intensity λ and $\Phi(l)$ denotes the locations of all active transmitters that transmit their packets in slot l.

In the above derivation we assume that the locations of nodes remain unchanged during the consecutive packet transmissions because the packet transmission times are relatively very short compared to the moving speed of a node.

To compute the expectation in the integrand of (2), let $BC_{\mathbf{x}}(k)$ denote the backoff counter value of the transmitter located at $\mathbf{x}$ in slot k. It then follows that

$$E\left[e^{-\mu\rho(\|\mathbf{z}\|)(1_{\{\mathbf{x}\in\Phi(0)\}}h_{\mathbf{x}}(0)+1_{\{\mathbf{x}\in\Phi(k)\}}h_{\mathbf{x}}(k))g(\|\mathbf{x}\|)}\right]$$

$$=P\{BC_{\mathbf{x}}(0)=1, BC_{\mathbf{x}}(k)=1\}E\left[e^{-\mu\rho(\|\mathbf{z}\|)(h_{\mathbf{x}}(0)g(\|\mathbf{x}\|)+h_{\mathbf{x}}(k)g(\|\mathbf{x}\|))}\right]$$

$$+ \sum_{j=2}^{W} P\{BC_{\mathbf{x}}(0) = 1, BC_{\mathbf{x}}(k) = j\} E\left[e^{-\mu\rho(||\mathbf{z}||)h_{\mathbf{x}}(0)g(||\mathbf{x}||)} \right]$$

$$+ \sum_{i=2}^{W} P\{BC_{\mathbf{x}}(0) = i, BC_{\mathbf{x}}(k) = 1\} E\left[e^{-\mu\rho(||\mathbf{z}||)h_{\mathbf{x}}(k)g(||\mathbf{x}||)} \right]$$

$$+ \sum_{i=2}^{W} \sum_{j=2}^{W} P\{BC_{\mathbf{x}}(0) = i, BC_{\mathbf{x}}(k) = j\}. \tag{3}$$

Let $\mathbf{q}$ be the stationary distribution of the backoff counter transition probability matrix $\mathbf{P}$, i.e., $\mathbf{qP} = \mathbf{q} = (q_1, q_2, \cdots, q_W)$. So using (3) and the assumption that the fading powers are independent and identically distributed yield

$$E\left[e^{-\mu\rho(||\mathbf{z}||)(1_{\{\mathbf{x}\in\Phi(0)\}}h_{\mathbf{x}}(0)g(||\mathbf{x}||)+1_{\{\mathbf{x}\in\Phi(k)\}}h_{\mathbf{x}}(k)g(||\mathbf{x}||))} \right]$$

$$= q_1(\mathbf{P}^k)_{11} \left(E\left[e^{-\mu\rho(||\mathbf{z}||)h_{\mathbf{x}}(0)g(||\mathbf{x}||)} \right] \right)^2$$

$$+ \sum_{j=2}^{W} q_1(\mathbf{P}^k)_{1j} E\left[e^{-\mu\rho(||\mathbf{z}||)h_{\mathbf{x}}(0)g(||\mathbf{x}||)} \right]$$

$$+ \sum_{i=2}^{W} q_i(\mathbf{P}^k)_{i1} E\left[e^{-\mu\rho(||\mathbf{z}||)h_{\mathbf{x}}(0)g(||\mathbf{x}||)} \right]$$

$$+ \sum_{i=2}^{W} \sum_{i=2}^{W} q_i(\mathbf{P}^k)_{ij}$$

$$=: f_0(\mathbf{z}, \mathbf{x}, k). \tag{4}$$

Combining (2) and (4) yields

$$E\left[e^{-\mu\rho(||\mathbf{z}||)(I(0)+I(k))} \right] = \exp\left(-\int_{\mathbb{R}^2} 1 - f_0(\mathbf{z}, \mathbf{x}, k)\Lambda(d\mathbf{x}) \right). \tag{5}$$

We now consider two consecutive transmissions, denoted by the first and second transmissions, from the tagged transmitter. Let $S_i, i = 1, 2$ denote the event that the i-th transmission is successful. Then, from (1) and (5) we have

$$P\{S_1 \cap S_2\}$$

$$= \sum_{k=1}^{W} P\left\{ \frac{h(0)g(||\mathbf{z}||)}{N(0) + I(0)} > \rho, \frac{h(k)g(||\mathbf{z}||)}{N(k) + I(k)} > \rho, BC = k \right\}$$

$$= \sum_{k=1}^{W} P\{BC = k\} E\left[e^{-\mu\rho(||\mathbf{z}||)N(0)} \right] E\left[e^{-\mu\rho(||\mathbf{z}||)N(k)} \right]$$

$$\times \exp\left(-\int_{\mathbb{R}^2} 1 - f_0(\mathbf{z}, \mathbf{x}, k)\Lambda(d\mathbf{x}) \right). \tag{6}$$

Using a similar argument as above, it is easy to show that

$$P\{S_1\} = P\{S_2\} = E\left[e^{-\mu\rho(||\mathbf{z}||)N(0)}\right]\exp\left(-q_1\int_{\mathbb{R}^2}1-f_1(\mathbf{z},\mathbf{x})\,\Lambda(d\mathbf{x})\right) \quad (7)$$

where

$$f_1(\mathbf{z},\mathbf{x}) := E\left[e^{-\mu\rho(||\mathbf{z}||)h_{\mathbf{x}}(0)g(||\mathbf{x}||)}\right].$$

So it immediately follows that

$$\begin{aligned}
&P\{S_2|S_1\}\\
&= \frac{P\{S_1\cap S_2\}}{P\{S_1\}}\\
&= \frac{\sum_{k=1}^{W}P\{BC=k\}E\left[e^{-\mu\rho(||\mathbf{z}||)N(k)}\right]\exp\left(-\int_{\mathbb{R}^2}1-f_0(\mathbf{z},\mathbf{x},k)\Lambda(d\mathbf{x})\right)}{\exp\left(-q_1\int_{\mathbb{R}^2}1-f_1(\mathbf{z},\mathbf{x})\,\Lambda(d\mathbf{x})\right)}.
\end{aligned} \quad (8)$$

Our next step is to derive (5) and (8) explicitly. To this end, we have to calculate the following exponent

$$\int_{\mathbb{R}^2}1-f_0(\mathbf{z},\mathbf{x},k)\Lambda(d\mathbf{x}). \quad (9)$$

From (4), we can write (9) as follows:

$$\begin{aligned}
&\int_{\mathbb{R}^2}1-f_0(\mathbf{z},\mathbf{x},k)\Lambda(d\mathbf{x})\\
&= \int_{\mathbb{R}^2}\Big(1-q_1(\mathbf{P}^k)_{11}f_1(\mathbf{z},\mathbf{x})^2 - q_1\sum_{j=2}^{W}(\mathbf{P}^k)_{1j}f_1(\mathbf{z},\mathbf{x})\\
&\quad -\sum_{i=2}^{W}q_i(\mathbf{P}^k)_{i1}f_1(\mathbf{z},\mathbf{x}) - \sum_{i=2}^{W}\sum_{j=2}^{W}q_i(\mathbf{P}^k)_{ij}\Big)\Lambda(d\mathbf{x})\\
&= q_1(\mathbf{P}^k)_{11}\int_{\mathbb{R}^2}(1-f_1(\mathbf{z},\mathbf{x})^2)\Lambda(d\mathbf{x})\\
&\quad + q_1\sum_{j=2}^{W}(\mathbf{P}^k)_{1j}\int_{\mathbb{R}^2}(1-f_1(\mathbf{z},\mathbf{x}))\Lambda(d\mathbf{x}) + \sum_{i=2}^{W}q_i(\mathbf{P}^k)_{i1}\int_{\mathbb{R}^2}(1-f_1(\mathbf{z},\mathbf{x}))\Lambda(d\mathbf{x}). \quad (10)
\end{aligned}$$

Since $\mathbf{q}$ is the stationary distribution, we have

$$q_1 = q_1\sum_{j=1}^{W}(\mathbf{P}^k)_{1j} = q_1(\mathbf{P}^k)_{11} + q_1\sum_{j=2}^{W}(\mathbf{P}^k)_{1j},$$

$$q_1 = \sum_{i=1}^{W}q_i(\mathbf{P}^k)_{i1} = q_1(\mathbf{P}^k)_{11} + \sum_{i=2}^{W}q_i(\mathbf{P}^k)_{i1}.$$

Combining the above and (10), we rewrite (9) as follows:

$$\int_{\mathbb{R}^2} 1 - f_0(\mathbf{z}, \mathbf{x}, k)\Lambda(d\mathbf{x})$$

$$= q_1(\mathbf{P}^k)_{11} \int_{\mathbb{R}^2} (1 - f_1(\mathbf{z}, \mathbf{x})^2) - (1 - f_1(\mathbf{z}, \mathbf{x})) - (1 - f_1(\mathbf{z}, \mathbf{x}))\Lambda(d\mathbf{x})$$

$$+ 2q_1 \int_{\mathbb{R}^2} (1 - f_1(\mathbf{z}, \mathbf{x}))\Lambda(d\mathbf{x})$$

$$= -q_1(\mathbf{P}^k)_{11} \int_{\mathbb{R}^2} (1 - f_1(\mathbf{z}, \mathbf{x}))^2 \Lambda(d\mathbf{x}) + 2q_1 \int_{\mathbb{R}^2} (1 - f_1(\mathbf{z}, \mathbf{x}))\Lambda(d\mathbf{x}). \quad (11)$$

Recall that fading powers are assumed to be independent and exponentially distributed with mean $\frac{1}{\mu}$. Then we have

$$f_1(\mathbf{z}, \mathbf{x}) = E\left[e^{-\mu\rho(||\mathbf{z}||)h_x(0)g(||\mathbf{x}||)}\right]$$

$$= \frac{\mu}{\mu + \mu\rho(||\mathbf{z}||)g(||\mathbf{x}||)}$$

$$= \frac{1}{\rho(||\mathbf{z}||)g(||\mathbf{x}||) + 1}$$

$$= \frac{1}{\rho||\mathbf{z}||^\alpha ||\mathbf{x}||^{-\alpha} + 1}.$$

Thus the integrand in (11) is given by

$$1 - f_1(\mathbf{z}, \mathbf{x}) = \frac{\rho||\mathbf{z}||^\alpha ||\mathbf{x}||^{-\alpha}}{\rho||\mathbf{z}||^\alpha ||\mathbf{x}||^{-\alpha} + 1}$$

and

$$\int_{\mathbb{R}^2} (1 - f_1(\mathbf{z}, \mathbf{x}))^2 \Lambda(d\mathbf{x})$$

$$= \int_{\mathbb{R}^2} \left(\frac{\rho||\mathbf{z}||^\alpha ||\mathbf{x}||^{-\alpha}}{\rho||\mathbf{z}||^\alpha ||\mathbf{x}||^{-\alpha} + 1}\right)^2 \Lambda(d\mathbf{x})$$

$$= \int_0^{2\pi} \int_0^\infty \left(\frac{\rho||\mathbf{z}||^\alpha s^{-\alpha}}{\rho||\mathbf{z}||^\alpha s^{-\alpha} + 1}\right)^2 \lambda s\, ds\, d\theta \quad \text{since } \Lambda \text{ is a measure with intensity} \lambda$$

$$= \int_0^\infty \left(\frac{\rho||\mathbf{z}||^\alpha s^{-\alpha}}{\rho||\mathbf{z}||^\alpha s^{-\alpha} + 1}\right)^2 2\lambda\pi s\, ds$$

$$= \int_0^\infty \left(\frac{\rho||\mathbf{z}||^\alpha}{\rho||\mathbf{z}||^\alpha + s^\alpha}\right)^2 2\lambda\pi s\, ds$$

$$= \int_0^\infty \left(\frac{\rho||\mathbf{z}||^\alpha}{\rho||\mathbf{z}||^\alpha + u}\right)^2 2\lambda\pi u^{\frac{2}{\alpha}-1} \frac{1}{\alpha} du \quad \text{where } u := s^\alpha$$

$$= \lambda\delta\pi\rho^2||\mathbf{z}||^{2\alpha} \int_0^\infty \frac{u^{\delta-1}}{(\rho||\mathbf{z}||^\alpha + u)^2} du \quad \text{where } \delta := \frac{2}{\alpha} < 1$$

$$= \lambda \delta \pi \rho^2 ||\mathbf{z}||^{2\alpha} (\rho ||\mathbf{z}||^{\alpha})^{\delta-2} \frac{\Gamma(2-\delta)\Gamma(\delta)}{\Gamma(2)} \quad \text{for} \quad \int_0^\infty \frac{u^{\delta-1}}{(\eta+u)^k} du = \eta^{\delta-k} \frac{\Gamma(k-\delta)\Gamma(\delta)}{\Gamma(k)}$$

$$= \lambda \delta \pi \rho^{\delta} ||\mathbf{z}||^2 \Gamma(2-\delta)\Gamma(\delta)$$

$$= (1-\delta)\lambda \pi \rho^{\delta} ||\mathbf{z}||^2 \Gamma(1-\delta)\Gamma(1+\delta). \tag{12}$$

Similarly, we obtain

$$\int_{\mathbb{R}^2} (1 - f_1(\mathbf{z},\mathbf{x})) \Lambda(d\mathbf{x})$$

$$= \int_{\mathbb{R}^2} \frac{\rho ||\mathbf{z}||^{\alpha} ||\mathbf{x}||^{-\alpha}}{\rho ||\mathbf{z}||^{\alpha} ||\mathbf{x}||^{-\alpha} + 1} \Lambda(d\mathbf{x})$$

$$= \int_0^\infty \frac{\rho ||\mathbf{z}||^{\alpha}}{\rho ||\mathbf{z}||^{\alpha} + s^{\alpha}} 2\lambda \pi s ds$$

$$= \int_0^\infty \frac{\rho ||\mathbf{z}||^{\alpha}}{\rho ||\mathbf{z}||^{\alpha} + u} \lambda \delta \pi u^{\delta-1} du$$

$$= \lambda \delta \pi \rho ||\mathbf{z}||^{\alpha} (\rho ||\mathbf{z}||^{\alpha})^{\delta-1} \frac{\Gamma(1-\delta)\Gamma(\delta)}{\Gamma(1)}$$

$$= \lambda \delta \pi \rho^{\delta} ||\mathbf{z}||^2 \Gamma(1-\delta)\Gamma(\delta)$$

$$= \lambda \pi \rho^{\delta} ||\mathbf{z}||^2 \Gamma(1-\delta)\Gamma(1+\delta). \tag{13}$$

Hence, we obtain (9) by combining (11), (12) and (13) as follows:

$$\int_{\mathbb{R}^2} 1 - f_0(\mathbf{z},\mathbf{x},k)\Lambda(d\mathbf{x})$$

$$= \left[-q_1(\mathbf{P}^k)_{11}(1-\delta) + 2q_1 \right] \lambda \pi \rho^{\delta} ||\mathbf{z}||^2 \Gamma(1-\delta)\Gamma(1+\delta).$$

Consequently,

$$E\left[e^{-\mu \rho(||\mathbf{z}||)(I(0)+I(k))} \right]$$

$$= \exp\left(-\int_{\mathbb{R}^2} 1 - f_0(\mathbf{z},\mathbf{x},k)\Lambda(d\mathbf{x}) \right)$$

$$= \exp\left(-q_1 \left[2 - (\mathbf{P}^k)_{11}(1-\delta) \right] \lambda \pi \rho^{\delta} ||\mathbf{z}||^2 \Gamma(1-\delta)\Gamma(1+\delta) \right). \tag{14}$$

We are now ready to provide our main theorem.
THEOREM 1

$$P\{S_1\} = E\left[e^{-\mu \rho(||\mathbf{z}||)N(0)} \right] \exp\left(-q_1 \lambda \pi \rho^{\delta} ||\mathbf{z}||^2 \Gamma(1-\delta)\Gamma(1+\delta) \right), \tag{15}$$

$$P\{S_2|S_1\} = \sum_{k=1}^W \left[P\{BC = k\}E\left[e^{-\mu \rho(||\mathbf{z}||)N(k)} \right] \right.$$

$$\left. \times \exp\left(-q_1 \left[1 - (\mathbf{P}^k)_{11}(1-\delta) \right] \lambda \pi \rho^{\delta} ||\mathbf{z}||^2 \Gamma(1-\delta)\Gamma(1+\delta) \right) \right]. \tag{16}$$

Proof. The success probability $P\{S_1\}$ is obtained by combining (7) and (13). The conditional probability $P\{S_2|S_1\}$ is also obtained from (8), (13) and (14).

Since we assume that the noise power process $\{N(k), k \geq 0\}$ forms independent and identically distributed random variables, we can rewrite (16) as follows:

$$
\begin{aligned}
P\{S_2|S_1\} \\
= E\left[e^{-\mu\rho(||\mathbf{z}||)N(0)}\right] &\exp\left(-q_1\lambda\pi\rho^\delta||\mathbf{z}||^2\Gamma(1-\delta)\Gamma(1+\delta)\right) \\
&\times \sum_{k=1}^{W} P\{BC=k\}\exp\left(q_1(\mathbf{P}^k)_{11}(1-\delta)\lambda\pi\rho^\delta||\mathbf{z}||^2\Gamma(1-\delta)\Gamma(1+\delta)\right) \\
= P\{S_1\} &\sum_{k=1}^{W} P\{BC=k\}\exp\left(q_1(\mathbf{P}^k)_{11}(1-\delta)\lambda\pi\rho^\delta||\mathbf{z}||^2\Gamma(1-\delta)\Gamma(1+\delta)\right) \\
= P\{S_1\}&\Psi(\lambda,||\mathbf{z}||,\alpha)
\end{aligned}
$$

where

$$
\begin{aligned}
\Psi(\lambda,||\mathbf{z}||,\alpha) \\
:= \sum_{k=1}^{W} P\{BC=k\}&\exp\left(q_1(\mathbf{P}^k)_{11}(1-\frac{2}{\alpha})\lambda\pi||\mathbf{z}||^2\rho^{\frac{2}{\alpha}}\Gamma(1-\frac{2}{\alpha})\Gamma(1+\frac{2}{\alpha})\right).
\end{aligned}
\tag{17}
$$

Note that $\Psi(\lambda,||\mathbf{z}||,\alpha)$ is the ratio of $P\{S_2|S_1\}$ over $P\{S_1\}$. So, if $\Psi(\lambda,||\mathbf{z}||,\alpha)$ were equal to 1, we might have the validity of the independence assumption on individual packet transmission results. However, since $\alpha > 2$, $\Psi(\lambda,||\mathbf{z}||,\alpha)$ is always greater than 1. Hence, we see that there obviously exists a correlation in the consecutive packet transmission results. However, if the ratio $\Psi(\lambda,||\mathbf{z}||,\alpha)$ is very close to 1, we might say that the independence assumption on consecutive packet transmission results is approximately valid. In the next section we call the ratio $\Psi(\lambda,||\mathbf{z}||,\alpha)$ by Dependency Factor (DF) and investigate the ratio $\Psi(\lambda,||\mathbf{z}||,\alpha)$ numerically to find the conditions where the independent assumption is approximately valid.

4   Numerical Results

In this section, we first provide numerical and simulation results to verify our analytical result on the DF in (17). We then plot the DF under various network scenarios to find the conditions where the independence assumption on consecutive packet transmissions is approximately valid.

In numerical and simulation studies, we assume that each node uses uniform distribution on $\{1, 2, \cdots, W\}$ to select its backoff counter value. For simulation we use MATLAB with parameters given in Table 1 and consider a ball-shaped

Table 1. Network parameters

λ	$0.01 \sim 0.15$ (points/m^2)
$\|\mathbf{z}\|$	$1 \sim 20$ (m)
α	$3, 4$
ρ	3.1623 (=5 dB)
μ	1
W	$63, 127$
R_S	60 (m)

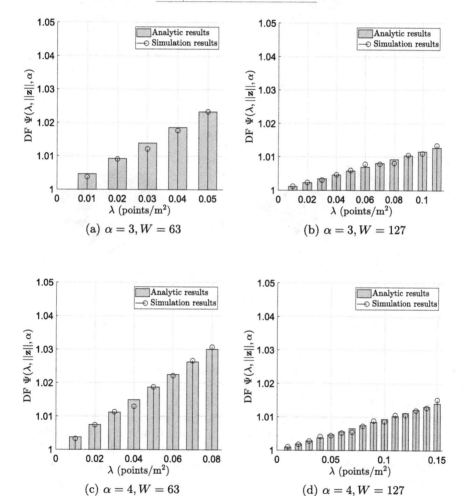

(a) $\alpha = 3, W = 63$

(b) $\alpha = 3, W = 127$

(c) $\alpha = 4, W = 63$

(d) $\alpha = 4, W = 127$

Fig. 1. Comparison of the analytic results (17) with simulation results where $\|\mathbf{z}\| = 10$ (m).

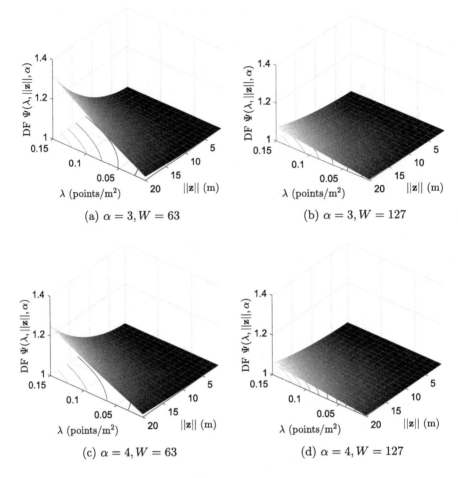

(a) $\alpha = 3, W = 63$

(b) $\alpha = 3, W = 127$

(c) $\alpha = 4, W = 63$

(d) $\alpha = 4, W = 127$

Fig. 2. Dependency Factor under various network parameters

region around the tagged receiver with the radius R_S. We assume that nodes are distributed in the region according to a HPPP with intensity λ. We also assume perfect sensing, i.e., every node has correct information on whether the channel is idle or not. The range of λ is restricted so that the probability of having a successful transmission is greater than 0.1 for given α and W. The results are plotted in Fig. 1. As seen in the figure, our analytical results of the DF in (17) are well matched with simulation results, which verifies that our analysis is valid.

It is worth noting that the DF $\Psi(\lambda, ||\mathbf{z}||, \alpha)$ is an increasing function of λ and $||\mathbf{z}||$; see (17) and Fig. 2. The reason for this is explained as follows. As λ and $||\mathbf{z}||$ increase, the network condition for the tagged receiver becomes worse and it becomes more difficult for the tagged receiver to have a successful packet reception, i.e., the tagged receiver has a smaller success probability $P\{S_1\}$. Therefore, once the tagged receiver has a successful reception in a network with large λ and

$||\mathbf{z}||$, it is reasonable to understand that the network condition becomes relatively good (for example, less contenders) for given λ and $||\mathbf{z}||$. Accordingly, the next packet transmission is also more likely to be successful. With a similar reason, we see that the DF decreases as W increases because increasing the window size W decreases the access opportunities of nodes in the network and hence eases competition between the nodes. We also see that the DF decreases as α increases because the increase in α results in that the number of significant interferers becomes less and this increases the success probability and decreases the DF.

A closer look at (15) and (17) gives the following interesting observations. Note that $\lambda\pi||\mathbf{z}||^2$ is the expected number of nodes closer to the tagged receiver than the tagged transmitter. So, we can interpret $M := \lambda\pi||\mathbf{z}||^2$ in (17) as the expected number of strong interferers to the tagged receiver. Note also that $P\{S_1\}$ and $\Psi(\lambda, ||\mathbf{z}||, \alpha)$ are functions of $\lambda\pi||\mathbf{z}||^2$. To see the impact of M on the DF and the success probability explicitly, we plot $\Psi(\lambda, ||\mathbf{z}||, \alpha)$ and $P\{S_1\}$ in Fig. 3 when we change the value of M. The figure shows that the higher the success probability, the lower the DF. Recall that the independent assumption on individual packet transmission results is approximately valid only when the DF is close to 1. From our observations above we conclude that the independent assumption is only valid for the network environment is relatively good, i.e., the expected number of nodes around the tagged receiver is relatively small. On the other hand, the independent assumption is not valid when the network environment is relatively bad, i.e., the expected number of nodes around the tagged receiver is relatively large. Therefore, it is necessary to develop a new mathematical model for the performance analysis of WLAN without the independence assumption for a super-dense network like a Small-World Super-Dense Device-to-Device wireless Network [18].

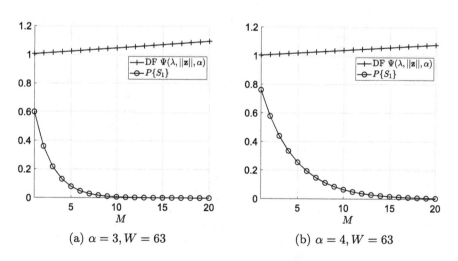

(a) $\alpha = 3, W = 63$ (b) $\alpha = 4, W = 63$

Fig. 3. $\Psi(\lambda, ||\mathbf{z}||, \alpha)$ versus $M := \lambda\pi||\mathbf{z}||^2$

5   Conclusions

We developed a mathematical model to consider the spatial distribution of nodes and the MAC protocol for channel access in WLAN. Based on our mathematical model we analyze the performance of WLAN and verify our analysis through numerical and simulation studies. From our analysis we investigate the performance behavior and the impact of the spatial distribution of node on the performance of WLAN.

Acknowledgment. This research was supported by Basic Science Research Program through the National Research Foundation of Korea (NRF) funded by the Ministry of Education (NRF-2017R1A2B4008581).

References

1. Bianchi, G.: Performance analysis of the IEEE 802.11 distributed coordination function. IEEE J. Sel. Areas Commun. **18**(3), 535–547 (2000)
2. Ye, S.R., Tseng, Y.C.: A multichain backoff mechanism for IEEE 802.11 WLANs. IEEE Trans. Veh. Technol. **55**(5), 1613–1620 (2006)
3. Kumar, A., Altman, E., Miorandi, D., Goyal, M.: New insights from a fixed-point analysis of single cell IEEE 802.11 WLANs. IEEE/ACM Trans. Netw. (TON) **15**(3), 588–601 (2007)
4. He, Y., Sun, J., Ma, X., Vasilakos, A.V., Yuan, R., Gong, W.: Semi-random backoff: towards resource reservation for channel access in wireless LANs. IEEE/ACM Trans. Netw. **21**(1), 204–217 (2013)
5. Toledo, A.L., Vercauteren, T., Wang, X.: Adaptive optimization of IEEE 802.11 DCF based on Bayesian estimation of the number of competing terminals. IEEE Trans. Mob. Comput. **5**(9), 1283–1296 (2006)
6. Chun, S., Xianhua, D., Pingyuan, L., Han, Z.: Adaptive access mechanism with optimal contention window based on node number estimation using multiple thresholds. IEEE Trans. Wirel. Commun. **11**(6), 2046–2055 (2012)
7. Oh, Y., Kim, Y., Hwang, G., Park, S.: A new contention based adaptive MAC protocol based on the renewal access protocol. In: 2016 IEEE 27th Annual International Symposium on Personal, Indoor, and Mobile Radio Communications (PIMRC), pp. 1–6. IEEE (2016)
8. Haenggi, M., Andrews, J.G., Baccelli, F., Dousse, O., Franceschetti, M.: Stochastic geometry and random graphs for the analysis and design of wireless networks. IEEE J. Sel. Areas Commun. **27**(7), 1029–1046 (2009)
9. ElSawy, H., Hossain, E., Haenggi, M.: Stochastic geometry for modeling, analysis, and design of multi-tier and cognitive cellular wireless networks: a survey. IEEE Commun. Surv. Tutor. **15**(3), 996–1019 (2013)
10. Bandyopadhyay, S., Coyle, E.J.: An energy efficient hierarchical clustering algorithm for wireless sensor networks. In: Twenty-Second Annual Joint Conference of the IEEE Computer and Communications, INFOCOM 2003, vol. 3, 1713–1723. IEEE Societies, IEEE (2003)
11. Andrews, J.G., Baccelli, F., Ganti, R.K.: A tractable approach to coverage and rate in cellular networks. IEEE Trans. Commun. **59**(11), 3122–3134 (2011)

12. Hu, Z., Lu, Z., Wen, X., Li, Q.: Stochastic-geometry-based performance analysis of delayed mobile data offloading with mobility prediction in dense IEEE 802.11 networks. IEEE Access **5**, 23060–23068 (2017)
13. Nguyen, H.Q., Baccelli, F., Kofman, D.: A stochastic geometry analysis of dense IEEE 802.11 networks. In: 26th IEEE International Conference on Computer Communications, INFOCOM 2007, pp. 1199–1207. IEEE (2007)
14. Kim, Y., Hwang, G.: Design and analysis of medium access protocol: throughput and short-term fairness perspective. IEEE/ACM Trans. Netw. (TON) **23**(3), 959–972 (2015)
15. Kim, Y., Hwang, G., Um, J., Yoo, S., Jung, H., Park, S.: Throughput performance optimization of super dense wireless networks with the renewal access protocol. IEEE Trans. Wirel. Commun. **15**(5), 3440–3452 (2016)
16. Kim, Y., Hwang, G.: Delay analysis and optimality of the renewal access protocol. Ann. Oper. Res. **252**(1), 41–62 (2017)
17. Haenggi, M.: Stochastic Geometry for Wireless Networks. Cambridge University Press, New York (2012)
18. Cheng, W., Yu, J., Zhao, F., Cheng, X.: Ssdnet: Small-world super-dense device-to-device wireless networks. IEEE Netw. **32**(1), 186–192 (2018)

Cognitive Radio Networks with Non-Ideal Spectrum Sensing and Multiple Classes of Secondary Users

Yuan Zhao[1]([✉]) and Wuyi Yue[2]

[1] School of Computer and Communication Engineering,
Northeastern University at Qinhuangdao, Qinhuangdao 066004, China
yuanzh85@163.com
[2] Department of Intelligence and Informatics,
Konan University, Kobe 658-8501, Japan

Abstract. In cognitive radio networks, spectrum sensing errors are unavoidable, and it is necessary to evaluate the negative effects of spectrum sensing errors on the system performance. In this paper, we consider and analyze the system performance of a cognitive radio network with non-ideal spectrum sensing. Considering the diversity of data transmission in modern communication networks, we also introduce multiple classes of secondary users (SUs) into the system evaluation. By building and analyzing a discrete-time Markov chain model, we obtain formulas for the collision probability, and the interruption rate and the average delay for the SU packets with lower priority. Finally, with the help of numerical results, we show the influence exerted the non-ideal spectrum sensing.

Keywords: Cognitive radio networks · Non-ideal spectrum sensing
Multiple classes of secondary users · Performance analysis

1 Introduction

Secondary users (SUs) have spectrum sensing abilities in cognitive radio networks. These SUs can sense the state of the spectrum and transmit opportunistically on the spectrum holes [1].

The spectrum sensing ability of the SUs allows them to avoid interfering with the transmission of the primary users (PUs). However, it has been posited that this ability may technically restricted, and as such, not ideal [2]. Consequently, in recent years, some researches, such as [3,4], have begun to explore the influence of non-ideal spectrum sensing on different performance measures in cognitive radio networks.

Most available research about non-ideal spectrum sensing does not consider a network environment with multiple classes of SUs. In modern networks, data transmission needs are many and varied. Therefore, in cognitive radio networks,

© Springer International Publishing AG, part of Springer Nature 2018
Y. Takahashi et al. (Eds.): QTNA 2018, LNCS 10932, pp. 160–167, 2018.
https://doi.org/10.1007/978-3-319-93736-6_12

it is necessary to introduce an SU grading mechanism to differentiate the diversity of data transmission needs among SUs. It is noteworthy that some previous studies have focused on grading the SUs in cognitive radio networks [5–8]. But all of these studies mentioned assumed ideal spectrum sensing abilities.

We note that the research that most relates to this paper is that done by [4,6] mentioned above. In [4], the non-ideal spectrum sensing results were introduced into a cognitive radio network with single-class SU packets. As a point of difference from [4], in this paper, we focus on multiple classes of SUs to capture the diversity of data transmission needs. In [6], a cognitive radio network with multiple classes of SUs but ideal spectrum sensing results was analyzed. Different from [6], in this paper, we introduce and evaluate the effects of non-ideal spectrum sensing results on the system performance of multiple classes of SUs.

Moreover, we note in cognitive radio systems, the SUs have different transmission demands. For example, one is real-time transmission demand (such as voice or video), which is strict with regard to instantaneous delivery but relatively tolerance of bit errors, and no-real-time transmission demand, which is strict with regard to bit errors but relatively tolerance of instantaneous delivery. We introduce two classes of SUs, SU1 and SU2 in this paper.

There are several innovative aspects to our paper. Firstly, we introduce simultaneously both non-ideal spectrum sensing and multiple classes of SUs into the performance evaluation of cognitive radio networks. Where the SU1 can be considered to be an SU with real-time transmission demands, and the SU2 can be considered to be an SU with non-real-time transmission demands. Secondly, different from the continuous-time models built into most existing research, in this paper, we overcome the complexity of discrete-time modeling analysis and build a discrete-time Markov chain model to perform the system performance analysis. With assumption of non-ideal spectrum sensing brought into consideration, we conduct numerical experiments to evaluate the influence of non-ideal spectrum sensing on different performance measures.

We organize the paper as follows. We present a discrete-time Markov chain model in Sect. 2 and analyze this Markov chain model in Sect. 3. We then derive the formulas for the collision probability, the SU2 packet interruption rate and the average delay and also show the numerical results for different performance measures in Sect. 4. Finally, we conclude our work in Sect. 5.

2   System Model

We focus on the system actions of the network users on a single channel. There is one PU in the system. The PU packets generated from the PU are endowed with the highest priority to occupy the channel. Considering the diversity of the SUs, we assume that there are two classed of SUs, SU1 and SU2 in the system. Therefore, in the system model considered in this paper, there is one PU and two classes of SUs, SU1 and SU2. Compared with PU packets and SU1 packets, SU2 packets have the lowest priority.

On the other hand, in real radio systems, the buffers can be employed to hold the packets. We assume that the SU2 has a buffer to reduce SU2 packets' loss. Correspondingly, no buffers are set to accommodate PU packets and SU1 packets.

Generally speaking, two types of non-ideal spectrum sensing results for SU packets should be considered: missed detection and false alarm [3]. We note that the non-ideal spectrum sensing results of SU2 packets will influence the transmission of both PU packets and SU1 packets. In this paper, we focus on evaluating the effect of spectrum sensing errors of the SU2 packets on the system performance. The two examples of sensing errors that we consider are false alarms and missed detections.

We build an early arrival system model, and consider a slotted time structure. We denote the slot boundaries as $t = 1, 2, \ldots$ and assume that the SU2 can sense the channel at the boundary of each slot. There are two possibilities for a false alarm. Firstly, an SU2 packet cannot access a channel that is in an idle state. Secondly, an SU2 packet can wrongly interrupt its ongoing transmission and return back to the SU2 buffer. In the case of missed detection, if a missed detection occurs, an SU2 packet will collide with a PU packet (or an SU1 packet) and then all the packets in the collision will be lost. We denote the notation $p_f(\bar{p}_f = 1 - p_f)$ as the false alarm rate and the notation $p_m(\bar{p}_m = 1 - p_m)$ as the missed detection rate.

With the same assumptions as in [4,6], the arrival processes of the three types of packets follow a Bernoulli process with arrival rates $\lambda_1(\bar{\lambda}_1 = 1 - \lambda_1)$, $\lambda_{21}(\bar{\lambda}_{21} = 1 - \lambda_{21})$ and $\lambda_{22}(\bar{\lambda}_{22} = 1 - \lambda_{22})$, respectively. The transmission times of the three types of packets follow geometric distributions with service rates $\mu_1(\bar{\mu}_1 = 1 - \mu_1)$, $\mu_{21}(\bar{\mu}_{21} = 1 - \mu_{21})$ and $\mu_{22}(\bar{\mu}_{22} = 1 - \mu_{22})$, respectively.

Let S_n be the number of SU2 packets and C_n be the channel state at the instant $t = n^+$. $C_n = 0$ denotes that the channel is not occupied by any packets. $C_n = 1, 2, 3$ denote that the channel is occupied by a PU packet, an SU1 packet or an SU2 packet, respectively. Specifically, $C_n = 4$ denotes the channel is in a state of confusion caused by packet collisions. Then $\{S_n, C_n\}$ constitutes a two-dimensional Markov chain. The state space $\boldsymbol{K}$ of $\{S_n, C_n\}$ can be given as follows:

$$\boldsymbol{K} = \{(0, j) : j = 0, 1, 2\} \cup \{(i, j) : i \geq 1, j = 0, 1, 2, 3, 4\}.$$

3   Model Analysis

Based on the possible state transitions of $\{S_n, C_n\}$, we define and discuss the block-structured transition probability matrix $\boldsymbol{P}$ as follows:

$$\boldsymbol{P} = \begin{pmatrix} \boldsymbol{U}_{00} & \boldsymbol{U}_{01} & & & \\ \boldsymbol{U}_{10} & \boldsymbol{X}_1 & \boldsymbol{X}_0 & & \\ & \boldsymbol{X}_2 & \boldsymbol{X}_1 & \boldsymbol{X}_0 & \\ & & \boldsymbol{X}_2 & \boldsymbol{X}_1 & \boldsymbol{X}_0 \\ & & & \ddots & \ddots & \ddots \end{pmatrix}. \tag{1}$$

Based on the following equations, we discuss each non-zero block in $\boldsymbol{P}$. We introduce notations $\rho = \lambda_1 \mu_1 + \bar{\mu}_1$, $\zeta = \lambda_{21} \mu_{21} + \bar{\mu}_{21}$ and $\vartheta = \lambda_{22} \mu_{22} + \bar{\lambda}_{22} \bar{\mu}_{22}$.

(1) $\boldsymbol{U}_{00}$ is a 3×3 matrix given as follows:

$$\boldsymbol{U}_{00} = \begin{pmatrix} \bar{\lambda}_{22}\bar{\lambda}_{1}\bar{\lambda}_{21} & \bar{\lambda}_{22}\lambda_{1} & \bar{\lambda}_{22}\bar{\lambda}_{1}\lambda_{21} \\ \bar{\lambda}_{22}\bar{\lambda}_{1}\bar{\lambda}_{21}\mu_{1} & \bar{\lambda}_{22}\rho & \bar{\lambda}_{22}\bar{\lambda}_{1}\lambda_{21}\mu_{1} \\ \bar{\lambda}_{22}\bar{\lambda}_{1}\bar{\lambda}_{21}\mu_{21} & \bar{\lambda}_{22}\lambda_{1} & \bar{\lambda}_{22}\bar{\lambda}_{1}\zeta \end{pmatrix}.$$

(2) $\boldsymbol{U}_{01}$ is a 3×5 matrix given as follows:

$$\boldsymbol{U}_{01} = \lambda_{22}\boldsymbol{V} \tag{2}$$

where $\boldsymbol{V}$ can be given as follows:

$$\boldsymbol{V} = \begin{pmatrix} \bar{\lambda}_{1}\bar{\lambda}_{21}p_{f} & \lambda_{1}\bar{p}_{m} & \bar{\lambda}_{1}\lambda_{21}\bar{p}_{m} & \bar{\lambda}_{1}\bar{\lambda}_{21}\bar{p}_{f} & (1-\bar{\lambda}_{1}\bar{\lambda}_{21})p_{m} \\ \bar{\lambda}_{1}\bar{\lambda}_{21}\mu_{1}p_{f} & \rho\bar{p}_{m} & \bar{\lambda}_{1}\lambda_{21}\mu_{1}\bar{p}_{m} & \bar{\lambda}_{1}\bar{\lambda}_{21}\mu_{1}\bar{p}_{f} & (1-\mu_{1}\bar{\lambda}_{1}\bar{\lambda}_{21})p_{m} \\ \bar{\lambda}_{1}\bar{\lambda}_{21}\mu_{21}p_{f} & \lambda_{1}\bar{p}_{m} & \bar{\lambda}_{1}\zeta\bar{p}_{m} & \bar{\lambda}_{1}\bar{\lambda}_{21}\mu_{21}\bar{p}_{f} & (1-\mu_{21}\bar{\lambda}_{1}\bar{\lambda}_{21})p_{m} \end{pmatrix}.$$

(3) $\boldsymbol{U}_{10}$ is a 5×3 matrix given as follows:

$$\boldsymbol{U}_{10} = \begin{pmatrix} 0 & 0 & 0 \\ 0 & 0 & 0 \\ 0 & 0 & 0 \\ \mu_{22}\bar{\lambda}_{1}\bar{\lambda}_{21}\bar{\lambda}_{22} & \mu_{22}\lambda_{1}\bar{\lambda}_{22} & \mu_{22}\bar{\lambda}_{1}\lambda_{21}\bar{\lambda}_{22} \\ \bar{\lambda}_{1}\bar{\lambda}_{21}\bar{\lambda}_{22} & \lambda_{1}\bar{\lambda}_{22} & \bar{\lambda}_{1}\lambda_{21}\bar{\lambda}_{22} \end{pmatrix}. \tag{3}$$

(4) $\boldsymbol{X}_{1}$ is a 5×5 matrix given as follows:

$$\boldsymbol{X}_{1} = \begin{pmatrix} \bar{\lambda}_{22} & 0 & 0 & 0 & 0 \\ 0 & \bar{\lambda}_{22} & 0 & 0 & 0 \\ 0 & 0 & \bar{\lambda}_{22} & 0 & 0 \\ 0 & 0 & 0 & \vartheta & 0 \\ 0 & 0 & 0 & 0 & \lambda_{22} \end{pmatrix} \times \boldsymbol{W} \tag{4}$$

where $\boldsymbol{W}$ can be given as follows:

$$\boldsymbol{W} = \begin{pmatrix} \bar{\lambda}_{1}\bar{\lambda}_{21}p_{f} & \lambda_{1}\bar{p}_{m} & \bar{\lambda}_{1}\lambda_{21}\bar{p}_{m} & \bar{\lambda}_{1}\bar{\lambda}_{21}\bar{p}_{f} & (1-\bar{\lambda}_{1}\bar{\lambda}_{21})p_{m} \\ \bar{\lambda}_{1}\bar{\lambda}_{21}\mu_{1}p_{f} & \rho\bar{p}_{m} & \bar{\lambda}_{1}\lambda_{21}\mu_{1}\bar{p}_{m} & \bar{\lambda}_{1}\bar{\lambda}_{21}\mu_{1}\bar{p}_{f} & (1-\mu_{1}\bar{\lambda}_{1}\bar{\lambda}_{21})p_{m} \\ \bar{\lambda}_{1}\bar{\lambda}_{21}\mu_{21}p_{f} & \lambda_{1}\bar{p}_{m} & \bar{\lambda}_{1}\zeta\bar{p}_{m} & \bar{\lambda}_{1}\bar{\lambda}_{21}\mu_{21}\bar{p}_{f} & (1-\mu_{21}\bar{\lambda}_{1}\bar{\lambda}_{21})p_{m} \\ \bar{\lambda}_{1}\bar{\lambda}_{21}p_{f} & \lambda_{1}\bar{p}_{m} & \bar{\lambda}_{1}\lambda_{21}\bar{p}_{m} & \bar{\lambda}_{1}\bar{\lambda}_{21}\bar{p}_{f} & (1-\bar{\lambda}_{1}\bar{\lambda}_{21})p_{m} \\ \bar{\lambda}_{1}\bar{\lambda}_{21}p_{f} & \lambda_{1}\bar{p}_{m} & \bar{\lambda}_{1}\lambda_{21}\bar{p}_{m} & \bar{\lambda}_{1}\bar{\lambda}_{21}\bar{p}_{f} & (1-\bar{\lambda}_{1}\bar{\lambda}_{21})p_{m} \end{pmatrix}.$$

(5) $\boldsymbol{X}_{0}$ is a 5×5 matrix given as follows:

$$\boldsymbol{X}_{0} = \lambda_{22}\boldsymbol{Y} \tag{5}$$

where $\boldsymbol{Y}$ can be given as follows:

$$\boldsymbol{Y} = \begin{pmatrix} \bar{\lambda}_{1}\bar{\lambda}_{21}p_{f} & \lambda_{1}\bar{p}_{m} & \bar{\lambda}_{1}\lambda_{21}\bar{p}_{m} & \bar{\lambda}_{1}\bar{\lambda}_{21}\bar{p}_{f} & (1-\bar{\lambda}_{1}\bar{\lambda}_{21})p_{m} \\ \bar{\lambda}_{1}\bar{\lambda}_{21}\mu_{1}p_{f} & \rho\bar{p}_{m} & \bar{\lambda}_{1}\lambda_{21}\mu_{1}\bar{p}_{m} & \bar{\lambda}_{1}\bar{\lambda}_{21}\mu_{1}\bar{p}_{f} & (1-\mu_{1}\bar{\lambda}_{1}\bar{\lambda}_{21})p_{m} \\ \bar{\lambda}_{1}\bar{\lambda}_{21}\mu_{21}p_{f} & \lambda_{1}\bar{p}_{m} & \bar{\lambda}_{1}\zeta\bar{p}_{m} & \bar{\lambda}_{1}\bar{\lambda}_{21}\mu_{21}\bar{p}_{f} & (1-\mu_{21}\bar{\lambda}_{1}\bar{\lambda}_{21})p_{m} \\ \bar{\mu}_{22}\bar{\lambda}_{1}\bar{\lambda}_{21}p_{f} & \bar{\mu}_{22}\lambda_{1}\bar{p}_{m} & \bar{\mu}_{22}\bar{\lambda}_{1}\lambda_{21}\bar{p}_{m} & \bar{\mu}_{22}\bar{\lambda}_{1}\bar{\lambda}_{21}\bar{p}_{f} & \bar{\mu}_{22}(1-\bar{\lambda}_{1}\bar{\lambda}_{21})p_{m} \\ 0 & 0 & 0 & 0 & 0 \end{pmatrix}.$$

(6) $\boldsymbol{X}_2$ is a 5×5 matrix given as follows:

$$\boldsymbol{X}_2 = \begin{pmatrix} 0\,0\,0 & 0 & 0 \\ 0\,0\,0 & 0 & 0 \\ 0\,0\,0 & 0 & 0 \\ 0\,0\,0 & \mu_{22}\bar{\lambda}_{22} & 0 \\ 0\,0\,0 & 0 & \bar{\lambda}_{22} \end{pmatrix} \times \boldsymbol{Z} \tag{6}$$

where $\boldsymbol{Z}$ can be given as follows:

$$\boldsymbol{Z} = \begin{pmatrix} 0 & 0 & 0 & 0 & 0 \\ 0 & 0 & 0 & 0 & 0 \\ 0 & 0 & 0 & 0 & 0 \\ \bar{\lambda}_1\bar{\lambda}_{21}p_f & \lambda_1\bar{p}_m & \bar{\lambda}_1\lambda_{21}\bar{p}_m & \bar{\lambda}_1\lambda_{21}\bar{p}_f & (1-\bar{\lambda}_1\bar{\lambda}_{21})p_m \\ \bar{\lambda}_1\bar{\lambda}_{21}p_f & \lambda_1\bar{p}_m & \bar{\lambda}_1\lambda_{21}\bar{p}_m & \bar{\lambda}_1\lambda_{21}\bar{p}_f & (1-\bar{\lambda}_1\bar{\lambda}_{21})p_m \end{pmatrix}.$$

We further define the steady-state distribution $\pi_{i,j}$ of $\{S_n, C_n\}$ as follows:

$$\pi_{i,j} = \lim_{n \to \infty} P\{S_n = i, C_n = j\}. \tag{7}$$

From the structure of $\boldsymbol{P}$, we conclude that the Markov chain $\{S_n, C_n\}$ follows a Quasi Birth and Death (QBD) process. With the matrix-geometric solution method introduced in [9], we can obtain $\pi_{i,j}$ defined in Eq. (7).

4 Performance Measures and Numerical Results

4.1 Performance Measures

We try to give the formulas for the collision probability, the SU2 packet interruption rate and average delay in this subsection.

An SU2 packet may collide with a PU packet or an SU1 packet because of non-ideal spectrum sensing. In the system considered in this paper, there may be a state of confusion caused by packet collisions. For this, we define a collision probability β as the probability for this state of confusion in this system. Therefore, the formula for β can be given as follows:

$$\beta = \sum_{i=1}^{\infty} \pi_{i,4}. \tag{8}$$

The SU2 packet interruption rate γ is the average number of SU2 packets whose transmissions are interrupted per slot. There are two possible cases. In the first case, the SU2 packet being transmitted correctly detects the arrivals of packets with higher priority, and then returns to the buffer. In the second case, the SU2 packet being transmitted on the channel false alarms the arrivals of packets with higher priority, and then returns to the buffer. So the formula for γ can be given as follows:

$$\gamma = \sum_{i=1}^{\infty} \pi_{i,3} \bar{\mu}_{22} \left[\left(1 - \bar{\lambda}_1 \bar{\lambda}_{21} \right) \bar{p}_m + \bar{\lambda}_1 \bar{\lambda}_{21} p_f \right]. \tag{9}$$

The SU2 packet average delay τ is defined as the average time that an SU2 packet stays in the system. Based on Little's equation [10], the formula for τ can be given as follows:

$$\tau = \frac{\sum_{i=1}^{\infty} \sum_{j=0}^{4} i \pi_{i,j}}{\lambda_{22}}. \tag{10}$$

4.2   Numerical Results

In this subsection, we show how the non-ideal spectrum sensing results influence the system performance by using numerical results. Without loss of generality, the service rates for the packets are set as $\mu_1 = \mu_{21} = \mu_{22} = 0.5$ in the numerical results.

Figure 1 shows how the SU2 packet interruption rate γ changes with respect to the PU packet arrival rate λ_1 when $\lambda_{21} = \lambda_{22} = 0.1$.

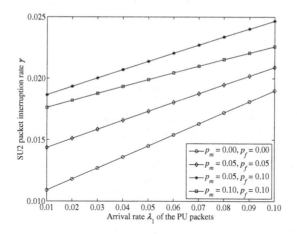

Fig. 1. SU2 packet interruption rate γ versus arrival rate λ_1.

In Fig. 1, the SU2 packet interruption rate γ is lower with the ideal spectrum sensing ($p_m = 0.00, p_f = 0.00$). That is to say, the non-ideal spectrum sensing results have detrimental effects on the system performance. Figure 1 also shows that a higher arrival rate λ_1 for PU packets can increase the SU2 packet interruption rate γ. Moreover, a higher false alarm rate p_f can increase the SU2 packet interruption rate γ. The reason is that more SU2 packets will interrupt their transmissions with a higher false alarm rate. Additionally, it's worth noting that as the missed detection rate p_m increases, the SU2 packet interruption rate γ will decrease. The reason may be that as the missed detection rate increases, more SU2 packets will leave the system because of packet collisions, and the possibility of the channel being idle will be higher. This obviously will lower the SU2 packet interruption rate.

Figure 2 shows the change trend for the SU2 packet average delay τ when $\lambda_1 = \lambda_{22} = 0.1$.

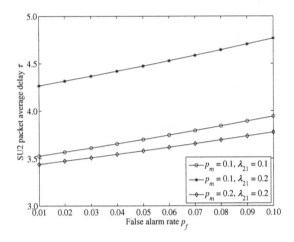

Fig. 2. SU2 packet average delay τ with false alarm rate p_f.

Figure 2 shows that as the false alarm rate p_f increases, the SU2 packet average delay τ will increase. This is because larger numbers of SU2 packets may be in the SU2 buffer under a higher false alarm rate. While the SU2 packet average delay τ can be decreased by setting a higher missed detection rate p_m. The reason for this interesting change trend may be that the higher the missed detection rate is, the more SU2 packets there are that will leave the system because of packet collisions, and then this will decrease the SU2 packet average delay. Moreover, Fig. 2 shows that a higher SU1 packet arrival rate λ_{21} also can cause an increase in the SU2 packet average delay τ. This is because the transmission probability for the SU2 packets is lower with a higher SU1 packet arrival rate.

From Figs. 1 and 2, we conclude that the SU2 packet interruption rate and the average delay will be increased under a higher false alarm rate, although these two performance measures seem to be decreased with a higher missed detection rate. But these results in fact come from larger numbers of packet collisions and losses.

5   Conclusion

Taking the non-ideal spectrum sensing results into consideration, this paper studied and evaluated the system performance of cognitive radio networks with two

Several approaches can be used to conserve energy and extend the sensor nodes lifetime [2, 3]. The duty cycling technique can save energy by putting nodes on sleep mode when there is no network activity. The radio is put to rest whenever possible in sleep mode (low-energy mode), and is brought back to active mode upon the occurrence of some event or, alternatively, by following some scheduling approach [4]. Additionally, the energy consumption of a sensor node during active mode can be more reduced by operating the radio at different frequency levels [5]. Thereby, we can save the amount of energy wasted in the transition frequency of the radio from the idle state to the busy state to start the transmission of data packets. This is an easy way to save the sensor energy by allowing it to be inactive (or on vacation) during idle state. during this period of time, it can only receive data packets. However, it can both receive and transmit data packets during busy state, and switches to idle state as soon as all queued packets are transmitted.

In the literature, vacation queueing models are widely used to model and analyze the dynamic behavior of sensor nodes with energy-saving. Almost works consider variants of N-policy vacation queue [6–11]. The N-policy is designed as an efficient queued wakeup scheme characterized by the fact that customers are accumulated and the idle server is turned on whenever there are N ($N \geq 1$) or more customers in the system, and that the server is turned off when the system becomes empty. The authors of [6, 7] investigated a variant of N-policy vacation queue with infinite buffer to minimize energy consumption for full-duplex nodes. Similarly, the papers [8, 9] considered a variant of limited capacity N-policy vacation queue based on half-duplex nodes for simplicity. In the same way, Nidehi and Goswami [10] considered a finite N-policy vacation queue for full-duplex nodes.

It has been proven in the literature that the N-policy is an efficient method for improving the lifetime of WSNs, but it can have a significant negative impact on the latency delay of data packets. To minimize the induced latency delay, Jiang et al. [12] proposed the Min(N,T) policy queue with infinite buffer for full-duplex nodes, where in addition to the queue threshold N, the radio can switch to the busy state when the waiting time of the leading packet has reached a fixed T time slots. On the other hand, based on full-duplex nodes with finite buffer, we have proposed in [11], a new vacation policy called Hybrid-policy, which can be viewed as a combination of the N-policy with the random vacation. Hence, a node can switch from idle state to busy state and starts the transmission of queued packets, if its buffer size reached the threshold N or at the end of the vacation period whose duration is random.

The aim of the present paper is to propose an energy saving and latency efficiency technique for full-duplex wireless sensor nodes based on a combination of normal vacation and working vacation policy. During a working vacation period [13], the server remains working at a lower service rate rather than completely stopping service during ordinary vacation period. Recently, in the papers [14, 15], the authors introduced the N-policy to return the server from the working vacation period to busy (or service) mode. The novelty of our investigation and with the aim of improving energy and minimizing latency, we propose in this paper a new vacation policy that we call *the two thresholds working vacation policy*, which corresponds to the N-policy with two different thresholds to switch the server between ordinary vacation, working vacation and service states. Hence, a node can switch from idle state (ordinary vacation state) to semi-busy state (working vacation state) and starts the transmission of queued packets at lower rate, if its buffer size reached the threshold N_2. While being in semi-busy state, a node can switch to busy state (service state) and start transmission of queued packets at normal rate, if its buffer size reached the threshold N_1 or return to idle state when the buffer empties. Moreover, we propose to model a full-duplex sensor node during its active mode by considering the limited capacity buffer, using the Generalized Stochastic Petri Nets (GSPNs) high level formalism.

The paper is organized as follows. In Sect. 2, we provide the model description of a sensor node, while in Sect. 3, we present a detailed description of the proposed GSPN models. Then, in Sect. 4, we develop the steady state performance indices and energy consumption of the given models. Several numerical results have been carried out using the efficient software tool GreatSPN in Sect. 5. Section 6 concludes the paper.

2   Mathematical Description of a Sensor Node with Vacation

In our paper, we consider a single full-duplex and finite buffer sensor node during its active period. First, we propose to model a sensor node as a vacation queueing system with N-policy, in which packets (relayed and sensed) arrive following a Poisson process with parameter λ and the service times are exponentially distributed with mean $1/\mu_1$. We assume that the sensor node can alternate between busy and idle states, which correspond to the service and vacation states of a vacation queueing model. During the busy state, the node can generate sensing data, receive and transmit data packets until the buffer empties. At this moment, it switches to idle state. While being in idle state, a node can generate sensing data, receive relayed data packets and store them in the buffer until the number of queued packets reaches the threshold N_1.

In a second time, we propose a new vacation policy for the sensor node, as a combination of normal vacation and working vacation queueing system, that we call *the two thresholds working vacation policy*, in which the normal service and the working vacation service times are exponentially distributed with mean $1/\mu_1$, and $1/\mu_2$, respectively. We assume that the sensor node can alternate between busy, semi-busy and idle states, which correspond to the service, working vacation and ordinary vacation states of a queueing model. During the busy state, the node transmits data packets until the buffer empties with mean $1/\mu_1$. At this moment, it switches to idle state and remains in this state till the buffer size reaches the threshold N_2 where instantly it switches to the semi-busy state. Within the semi-busy state, packets are transmitted with mean $1/\mu_2$ ($\mu_2 < \mu_1$). If during the semi-busy state, the number of waiting packets reaches the threshold N_1, the sensor node immediately switches to the busy state or returns to idle state if the buffer becomes empty.

3   GSPN Models of a Sensor Node with the Two Vacation Policies

The Generalized Stochastic Petri Nets (GSPNs) [16] are a powerful mathematical and graphical formalism which allows to model easily the system behavior and to obtain interesting steady state performance indices of the modeled system. A GSPN consists mainly of a set of places, a set of transitions and directed arcs. Places are drawn as circles and may contain any number of tokens (black dots) which model the various resources of the system. Transitions represent actions and can be immediate transitions or timed transitions, which are drawn as thin lines and rectangles respectively. Timed transitions model the execution of time consuming actions and fire with a random, exponentially distributed firing delay, whereas immediate transitions model logic activities, like synchronization, fire without any delay and have priority over timed transitions. Places and transitions are linked by arcs which may be labeled by a multiplicity. However, the presence of tokens in places connected with inhibitor arcs to a transition (circle-headed arcs from places to transitions) inhibit the firing of this transition. The firing of a transition removes tokens from its input places (the number of tokens indicated on input arcs of the transition) and putting tokens to its output places (the number of tokens indicated on output arcs of the transition). The GSPN marking M is defined by the number of tokens contained in each place which represents the system state. M_0 is called the initial marking of a GSPN and determines its initial state.

In the following, we describe the proposed GSPNs models for wireless sensor node with the two vacation policies, namely N-policy and the two thresholds working vacation policy.

3.1   The GSPN Model with N-Policy

In Fig. 1, the wireless sensor node model with the N-policy queueing vacation discipline is graphically represented as a GSPN model.

- The place $Capacity$: represents the available storage capacity in the buffer of the sensor node. The buffer capacity is K packets represented by K tokens initially residing in the place $Capacity$.
- The place $Buffer$: represents the number of packets in the buffer waiting for transmission.
- The place $Idle$: represents that the radio is in idle state.
- The place $Busy$: represents that the radio is in busy state.
- The initial marking of the net is:
  $M0 = \{M(Capacity), M(Buffer), M(Idle), M(Busy)\} = \{K, 0, 1, 0\}$,
  which represents the fact that no packet is present in the node, the radio is idle and the buffer is empty.

Capacity

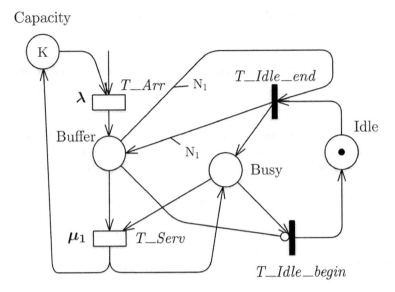

Fig. 1. GSPN of a sensor node with N-policy

- The firing of the timed transition T_Arr indicates the arrival of new packets (relayed or sensed).
- At the arrival of a new packet, if the place *Capacity* contains at least one token, which represents the fact that there is at least one idle position in the buffer, the transition T_Arr fires and produces one token in the place *Buffer*.
- Whenever the place *Busy* contains a token (The radio is in busy state) and the place *Buffer* is not empty, the timed transition T_Serv fires with mean rate μ_1 by removing one token from each input places (*Buffer* and *Busy*), remitting a token to the place *Busy* representing that the radio becomes ready to transmit another packet and producing another token in *Capacity* representing thus the liberation of a position in the buffer. This firing represents the transmission of a packets.
- The firing of the immediate transition T_Idle_begin represents the transition from busy to idle state. When the radio is in idle state (represented by a token in the place *Idle*) and the buffer is empty, which is verified by an inhibitor arc from the place *Buffer* to the immediate transition T_Idle_begin.
- The passage of the sensor node from the idle to the busy state is controlled by the N-policy. Hence, the end of the idle state (vacation period) corresponds to the firing of the immediate transition T_Idle_end as soon as the number of token reaches N_1 in the place *Buffer*, which corresponds to have N_1 waiting packets in the sensor node buffer.

3.2   The GSPN Model with the Two Thresholds Working Vacation Policy

The GSPN model with the two thresholds working vacation policy is depicted in Fig. 2. In this model, the passage from idle state to semi-busy state and from semi-busy state to busy state are both controlled by the N-policy but with different thresholds values (N_1 and N_2). To this aim, we add to the previous model (given in Fig. 1) the place *SemiBusy* which represents that the radio is in semi-busy state, the timed transition T_Serv_SB, two other immediate transitions T_SB_end and $T\_Idle\_begin_2$ and we rename the transition T_Idle_begin to $T\_Idle\_begin_1$. The initial marking of the net is $M0 = \{M(Capacity), M(Buffer), M(Idle), M(SemiBusy), M(Busy)\} = \{K, 0, 1, 0, 0\}$, which represents the fact that the buffer is initially empty and the radio is idle. When the sensor node is idle (place *Idle* contains a token) and after the arrival of the N_2th packet, the immediate transition T_Idle_end fires and removes a token from the place *Idle* and deposits a token in the place *SemiBusy*. This firing represents the transition from idle state to semi-busy state. While being in semi-busy state, the radio transmits packets using the lower mean rate μ_2 to reduce the latency and minimize the energy consumption by operating with a lower speed. The transmission during semi-busy state (working vacation) is achieved by the firing of the timed transition T_Serv_SB. Meanwhile, if the number of queued packets reaches the threshold N_1, the radio come back to the normal working level immediately (busy state) by firing the immediate transition T_SB_end. At a transmission completion of a packet during busy (semi-busy) state and the buffer is empty, the sensor node switches to idle state by the firing of the immediate transition $T\_S\_Idle\_begin$.

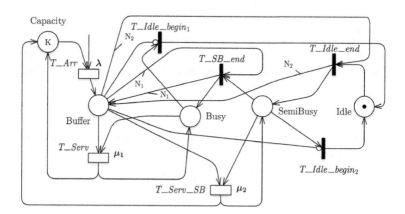

Fig. 2. GSPN of a sensor node with two thresholds working vacation policy

4   Performance Indices

As the GSPN models described in Sect. 3 are bounded and their initial marking M_0 is a home state, they are ergodic and the steady-state distribution exists and is unique. Hence, the stationary probability vector $\pi = (\pi_1, \pi_2, ..., \pi_n)$ can be computed and we can derive the exact steady-state performances indices and energy consumption of a wireless sensor node. The GreatSPN [17] software system is used to validate the correctness (qualitative analysis) of the developed GSPN models, compute the vector π and the performance measures using the formulas given below. In what follows, $M_i(p)$ indicates the number of tokens in place p in the marking M_i, π_i is the probability that the process is in the state M_i at steady state, A is the set of reachable tangible markings, and $E(t)$ is the set of tangible markings reachable by transition t.

- The blocking probability of packets (P_B): It corresponds to the buffer saturation probability.

$$P_B = \sum_{i:M_i(Capacity)=0} \pi_i. \tag{1}$$

- The probability that a sensor node is on idle state P_I: It corresponds to the probability that the place $Idle$ contains one token.

$$P_I = \sum_{i:M_i(Idle)=1} \pi_i. \tag{2}$$

- The Probability that a sensor node is on semi-busy state P_{SB}: It corresponds to the probability that the place $SemiBusy$ contains one token.

$$P_{SB} = \sum_{i:M_i(SemiBusy)=1} \pi_i. \tag{3}$$

- The mean number of packets in a sensor node $(\overline{Q})$: It represents the mean number of waiting packets in the sensor including the packet being transmitted. This corresponds to the mean number of tokens in the place $Buffer$.

$$\overline{Q} = \sum_{i:M_i \in A} M_i(Buffer) \cdot \pi_i. \tag{4}$$

- The packets reception throughput $\overline{\lambda}$: It corresponds to the effective rate of packets reception by the sensor node.

$$\overline{\lambda} = \lambda \cdot \sum_{i:M_i \in E(T\_Arr)} \pi_i. \tag{5}$$

- The average length of an idle period $\overline{I}$: It corresponds to the average duration of time when the place $Idle$ contains one token.

$$\overline{I} = \begin{cases} \frac{N_1}{\lambda} & \text{N-policy} \\ \\ \frac{N_2}{\lambda} & \text{Two thresholds working vacation policy} \end{cases}. \tag{6}$$

– The average length of a semi-busy period $\overline{SB}$: It corresponds to the average duration of time when the place $SemiBusy$ contains one token (in the second model), which represents the average time of packets transmission with low rate μ_2.

$$\overline{SB} = \min\{\frac{N_1 - N_2}{\lambda}, \frac{\overline{Q}_{SB}}{\mu_2}\}. \tag{7}$$

Where: $\overline{Q}_{SB} = \sum_{i:M_i \in A \& M_i(SemiBusy)=1} M_i(Buffer) \cdot \pi_i$ which represents the mean number of packets in the sensor node during semi-busy state.

– The average length of a busy period $\overline{B}$: It corresponds to the average duration of time when the radio transmits packets with high rate μ_1.

$$\overline{B} = \frac{\overline{Q}_B}{\mu_1}. \tag{8}$$

Where: $\overline{Q}_B = \sum_{i:M_i \in A \& M_i(Busy)=1} M_i(Buffer) \cdot \pi_i$ which represents the mean number of packets in the sensor node during busy state.

– The Average duration of a cycle $(\overline{C})$: From (6), (7) and (8), we can get the mean duration of one cycle per time unit.

$$\overline{C} = \begin{cases} \overline{B} + \overline{I} & \text{N-policy} \\ \overline{B} + \overline{SB} + \overline{I} & \text{Two thresholds working vacation policy} \end{cases} \tag{9}$$

– The mean sojourn time of packets $\overline{W}$: Using Little's formula, the expected mean sojourn time of packets in the sensor node is given by :

$$\overline{W} = \frac{\overline{Q}}{\lambda}. \tag{10}$$

– The energy consumption at a sensor node (EC): Based on Jiang et al. [7], we propose the following mean energy consumption formulas.

$$EC = \begin{cases} EC_I \cdot P_I + EC_b \cdot (1 - P_I) + \frac{EC_s}{\overline{C}} + \overline{Q} \cdot EC_{Tx} \\ \text{N-policy} \\ \\ EC_{SB} \cdot P_{SB} + EC_I \cdot P_I + EC_b \cdot (1 - P_{SB} - P_I) + \\ \frac{EC_s}{\overline{C}} + \overline{Q} \cdot EC_{Tx} \\ \text{Two thresholds working vacation policy} \end{cases} \cdot \tag{11}$$

where :

EC_{SB} energy consumption while the radio is in semi-busy state.

EC_I energy consumption while the radio is in idle state.

EC_b energy consumption while the radio is in busy state.

EC_{Tx} energy consumption for holding each packet present in the sensor node.

EC_s energy dissipated when a sensor node switches between the different states.

Table 1. System parameters

Parameter	Value
Capacity of buffer (K)	20
Queue threshold N_1	Range from 1 to $K-1$
Queue threshold N_2	2
Mean data arrival rate (λ)	Range from 0.25 to 0.75
Mean normal service rate (μ_1)	5
Mean working vacation service rate (μ_2)	Range from 0.25 to 0.75
EC_I	10
EC_{SB}	Range from 25 to 75
EC_b	500
EC_{Tx}	5
EC_s	300

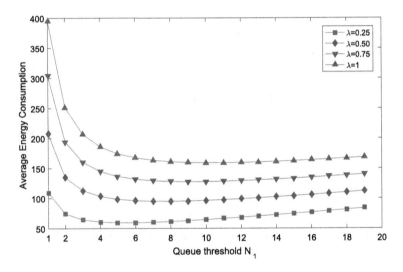

Fig. 3. Average energy consumption vs. queue threshold N_1, N-policy $\lambda : 0.25 \sim 1$

5   Experimental Results and Discussions

In this section, some numerical results are presented to show the efficiency of
the proposed models. The numerical results were computed using the GreatSPN
tool [17]. The system parameters are taken as shown in Table 1.

First, we investigate the effect of varying the queue threshold N_1 on the
average energy consumption for various λ values, when μ_1 is set to 5. Figure 3
considers the sensor node model with N-policy. It shows that the mean energy
consumption decreases when the queue threshold N_1 increases till it reaches an
optimal value of N_1 which mitigates the average energy consumption and then

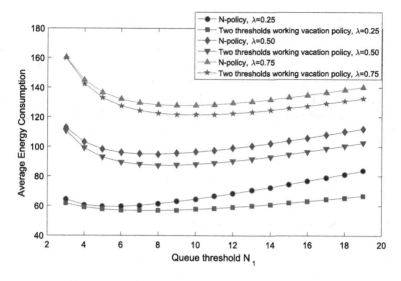

Fig. 4. Average energy consumption vs. queue threshold N_1, $\lambda : 0.25 \sim 0.75$

it increases thereafter. For example, the optimal value of N_1 can be reach at $N_1 = 5$ for $\lambda = 0.25$ and we can conserve until $45,73\%$ compared to the mean energy consumption for an ordinary sensor node ($N = 1$), which agree with the intuitive expectations. In Fig. 4, we we compare the average energy consumption for the N-policy and the two thresholds working vacation policy models, where the queue threshold during semi-busy state N_2 is set to 2 and μ_2 is set to 0.5. We notice that with the two thresholds working vacation policy, the average energy consumption also decreases as the queue threshold N_1 increases until it reaches an optimal value of N_1 which minimizes the average energy consumption and then it increases slowly thereafter. It can be seen that the average energy consumption for the model with two thresholds working vacation policy is less than the model with N-policy. Note that, within the sensor node model with the two thresholds working vacation policy, we can conserve for example over then 5.8% when $N_1 = 7$ and $\lambda = 0.25$. However, we can also conserve until 8% when $N_1 = 8$ and $\lambda = 0.50$.

Likewise we compare and examine the influence of the queue threshold N_1 on the mean sojourn time $\overline{W}$ of the proposed models (N-policy and Two thresholds working vacation policy). We assume that the queue threshold during semi-busy state N_2 is set to 2, μ_1 is set to 5 and μ_2 is set to 0.25. Figure 5 shows that $\overline{W}$ increases with the increasing of N_1. However, we can notice that $\overline{W}$ decreases with the increase of λ. This result agrees with the intuition, that N_1 is reached with the higher values of λ faster than the lower values. It can be also clearly seen that among the two models, the model with the two thresholds working vacation policy has the lowest $\overline{W}$, due to the fact that the radio remains transmitting data packets at a lower mean service rate μ_2 during semi-busy state.

Then, similarly we study the impact of the queue threshold N_1 on the blocking probability P_B for several values of λ when μ_1 is set to 5, $\mu_2 = 0.25$ and $N_2 = 2$. As shown in Fig. 6, the P_B increases by increasing N_1. We can also notice that among the two models, the model with two thresholds working vacation policy gives the best (lowest) P_B, This is due to the fact that the radio continues to transmit data packets at a lower service rate μ_2 during semi-busy state rather than completely stopping service in the model with N-policy.

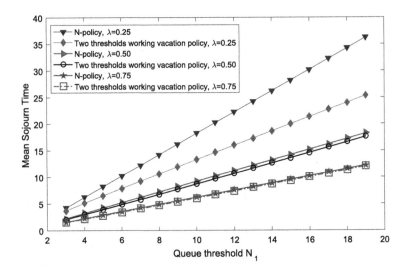

Fig. 5. Mean sojourn time vs. queue threshold N_1, $\lambda : 0.25 \sim 0.75$

Finally, we consider the model with two thresholds working vacation policy where we pay attention to the relation between the queue threshold N_1, the average energy consumption and the mean sojourn time for various semi-busy state mean service rate μ_2, when the mean arrival rate λ is set to 0.5, the normal mean service rate μ_1 is set to 5 and the queue threshold N_2 is set to 2. Figure 7 shows that the average energy consumption decreases with the increase of N_1 until it reaches an optimal queue threshold value of N_1 then it increases thereafter, however, the average energy consumption increases with the increase of μ_2. This is due to the fact that with the increase of μ_2, the EC_{SB} will also increase, which increases the average energy consumption. Figure 8 highlights the effects of N_1 and μ_2 on the mean sojourn time $\overline{W}$. It can be see that $\overline{W}$ decreases with the increase of N_1 while it decreases with the increase of μ_2. This results are logical because the increasing of μ_2 describes the fact that the radio will work faster, hence, the average energy consumption increases and the mean sojourn time $\overline{W}$ decreases.

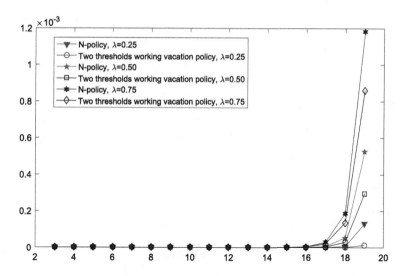

Fig. 6. Blocking probability P_B vs. queue threshold N_1, $\lambda : 0.25 \sim 0.75$

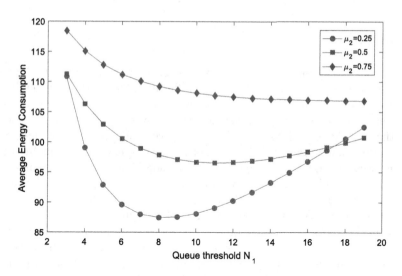

Fig. 7. Average energy consumption vs. queue threshold N_1, two thresholds working vacation policy $\mu_2 : 0.25 \sim 0.75$

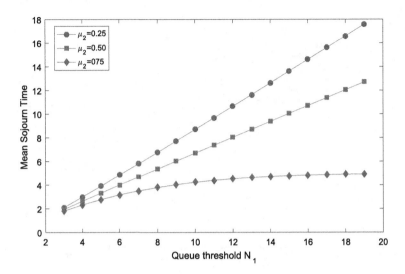

Fig. 8. Mean sojourn time vs. queue threshold N_1, two thresholds working vacation policy $\mu_2 : 0.25 \sim 0.75$

6 Conclusion

In this paper, we have proposed a new energy saving and latency efficiency technique for wireless sensor nodes. First, a finite buffer vacation queueing system with N-policy is used to model a sensor node. Then, a finite buffer vacation queueing system with two thresholds working vacation policy is considered. Using the high level GSPNs formalism, we have modeled a sensor node with each vacation policy and we have developed the formulas of the main steady state performance indices and energy consumption. The numerical results comparing the two models are made by using the efficient software tool GreatSPN and demonstrate that, although the sensor node model with N-policy reduces the energy consumption, the sensor node model with the two thresholds working vacation policy reduces more energy and has also a significant effect on minimizing the latency delay and the blocking probability. These results prove that the proposed sensor node model with two thresholds working vacation policy is an efficient technique for extending the lifetime of WSNs.

References

1. Akyildiz, I.F., Su, W., Sankarasubramaniam, Y., Cayirci, E.: Wireless sensor networks: a survey. Comput. Netw. **38**(4), 393–422 (2002)
2. Abdelaal, M., Theel, O. : Recent energy-preservation endeavours for longlife wireless sensor networks: a concise survey. In: The Eleventh International Conference on Wireless and Optical Communications Networks (WOCN 2014), pp. 1–7 (2014)

3. Abdul-Salaam, G., Abdullah, A.H., Anisi, M.H., Gani, A., Alelaiwi, A.: A comparative analysis of energy conservation approaches in hybrid wireless sensor networks data collection protocols. Telecommun. Syst. **61**(1), 159–179 (2016)
4. Wang, Y., Chen, H., Wu, X., Shu, L.: An energy-efficient SDN based sleep scheduling algorithm for WSNs. Netw. Comput. Appl. **59**, 39–45 (2016)
5. Pughat, A., Sharma, V.: A review on stochastic approach for dynamic power management in wireless sensor networks. Hum. Centric Comput. Inf. Sci. **5**(1), 1–14 (2015)
6. Jiang, F.C., Huang, D.C., Yang, C.T., Wang, K.H.: Mitigation techniques for the energy hole problem in sensor networks using N-policy M/G/1 queuing models. In: The International Conference on Frontier Computing, Theory, Technologies and Applications (IET 2010), pp. 281–286 (2010)
7. Jiang, F.C., Huang, D.C., Yang, C.T., Leu, F.Y.: Lifetime elongation for wireless sensor network using queue-based approaches. Supercomputing **59**(3), 1312–1335 (2012)
8. Huang, D.C., Tseng, H.C., Deng, D.J., Chao, H.C.: A queue-based prolong lifetime methods for wireless sensor node. Comput. Commun. **35**(9), 1098–1106 (2012)
9. Huang, D.C., Lee, J.H.: A dynamic N threshold prolong lifetime method for wireless sensor nodes. Math. Comput. Model. **57**(11), 2731–2741 (2013)
10. Nidhi, M., Goswami, V.: A randomized *N*-policy queueing method to prolong lifetime of wireless sensor networks. In: Nagar, A., Mohapatra, D.P., Chaki, N. (eds.) Proceedings of 3rd International Conference on Advanced Computing, Networking and Informatics. SIST, vol. 43, pp. 347–357. Springer, New Delhi (2016). https://doi.org/10.1007/978-81-322-2538-6_36
11. Boutoumi, B., Gharbi, N.: An energy saving and latency delay efficiency scheme for wireless sensor networks based on GSPNs. In: The 4th International Conference on Control, Decision and Information Technologies (CoDIT 2017), pp. 645–650 (2017)
12. Jiang, F.C., Huang, D.C., Yang, C.T., Lin, C.H., Wang, K.H.: Design strategy for optimizing power consumption of sensor node with Min (N, T) policy M/G/1 queuing models. Int. J. Commun. Syst. **25**(5), 652–671 (2012)
13. Servi, L.D., Finn, S.G.: M/M/1 queues with working vacations (M/M/1/WV). Perform. Eval. **50**(1), 41–52 (2002)
14. Sreenivasan, C., Chakravarthy, S.R., Krishnamoorthy, A.: MAP/PH/1 queue with working vacations, vacation interruptions and N-policy. Appl. Math. Model. **37**(6), 3879–3893 (2013)
15. Yang, D.Y., Wu, C.H.: Cost-minimization analysis of a working vacation queue with N-policy and server breakdowns. Comput. Ind. Eng. **82**, 151–158 (2015)
16. Marsan, M.A., Balbo, G., Conte, G., Donatelli, S., Franceschinis, G.: Modelling with Generalized Stochastic Petri Nets. John Wiley & Sons Inc., Hoboken (1994)
17. GRaphical Editor and Analyzer for Timed and Stochastic Petri Nets. http://www.di.unito.it/greatspn/index.html

Performance Analysis of a Dynamic Channel Vacation Scheme in Cognitive Radio Networks

Wenbo Wang[1], Zhanyou Ma[1(✉)], Wuyi Yue[2], and Yutaka Takahashi[3]

[1] School of Science, Yanshan University, Qinhuangdao 066004, China
wangwenbo931118@163.com, mzhy55@ysu.edu.cn
[2] Department of Intelligence and Informatics, Konan University,
Kobe 658-8501, Japan
yue@konan-u.ac.jp
[3] Graduate School of Informatics, Kyoto University, Kyoto 606-8225, Japan
takahashi@i.kyoto-u.ac.jp

Abstract. In order to resolve the issues of channel scarcity and low channel utilization rates in cognitive radio networks (CRNs), some researchers proposed to the idea of "secondary utilization" for the licensed channels. In "secondary utilization", secondary users (SUs) opportunistically take advantage of unused licensed channels, thus guaranteeing the transmission performance and quality of service (QoS) of the system. Based on the channel vacation scheme, we analyze a preemptive priority queueing system with multiple synchronization working vacations. Under this discipline, we build a three-dimensional Markov process for this queueing model. Through the analysis of performance measures, we obtain the average queue length of two types of users, the mean busy period and the channel utility. Finally, by analyzing several numerical experiments, we demonstrate the effect of the parameters on the performance measures.

Keywords: Channel vacation scheme · Three-dimensional markov process
Preemptive priority · Queueing model

1 Introduction

In the traditional static channel allocation method, the system allocates a fixed wireless channel for each server. Even though the use of this method had enabled the elimination of interference between different servers, it has also led to a high imbalance in channel utilization. Research reports show that licensed channel utilization ranges from 15% to 85%, and that the utilization of most licensed channels is very low [1]. However, remaining unlicensed channels are used frequently, which leads to an enormous waste of channel resources, and an increased shortage in wireless channel availability. How to take advantage of channel resources more efficiently has thus become a research focus. Cognitive radio networks (CRNs) have altered the thinking around traditional static channel allocation, and have given rise to the concept of dynamic channel distribution.

In general, there are two types of users and multiple licensed channels in CRNs. Primary Users (PUs) own the exclusive rights to licensed channels, and Secondary

© Springer International Publishing AG, part of Springer Nature 2018
Y. Takahashi et al. (Eds.): QTNA 2018, LNCS 10932, pp. 182–190, 2018.
https://doi.org/10.1007/978-3-319-93736-6_14

Users (SUs) own the ability to sense channels and opportunistically utilize unused licensed channels, which leads to the idea of "secondary utilization" for the licensed channels. Many researchers have investigated the multiple channels model using different strategies. For instance, Yu et al. considered a multi-channel CRN where each SU could only choose to sense a subset of channels [2].

Other researchers have concentrated on saving network resources that are wastefully depleted by idle channels. For instance, Zhao et al. investigated a kind of spectrum allocation strategy with dynamic channel closing scheme, in which parts of the channels could be closed periodically to realize the data transmission control once there are not any transmission requests in the system [3]. Wu et al. proposed an optimal resource allocation scheme in a multi-channel CRN, and considered the dynamic channel model and the sensing errors based on a vacation queueing model [4].

Additionally, different strategies using queueing models have been applied in CRNs. For example, Kim considered preemptive priority in CRNs by building a new T-preemptive priority M/G/1 queueing model, and calculated the average waiting time for PUs and SUs [5].

In our model, we mainly present both channel utility and resource conservation in cognitive radio networks with a dynamic multichannel vacation scheme controlled by central controller, and establish a Markov process to evaluate the system performance measures. We would like to consider the performance analyses of the wireless communication networks with interference or sensing protocol.

The remainder of this paper is organized as follows. In Sect. 2, the novel strategy and the queueing model we studied is described. In Sect. 3, the steady-state distribution of the queue length is analyzed. In Sect. 4, some performance measures are considered. In Sect. 5, some numerical experiments are given. In Sect. 6, conclusions are stated.

2 Channel Allocation Strategy and System Model

2.1 The Dynamic Channel Vacation Scheme

In this paper, we consider a CRN consisting of two types of users and multiple licensed channels. The central controller and web users interact with information about the licensed channels. In order to conserve the network resource properly, a kind of channel allocation strategy with a dynamic channel vacation scheme is proposed. In this strategy, all c channels can enter into a working vacation period to realize data transmission control once there are no remaining transmission requests in the system. The specific description for the CRN with a dynamic channel vacation scheme is shown as Fig. 1.

According to the dynamic channel vacation scheme, we can divide the system state into three states as follows:

1. Working vacation state. If the channels are empty, the channels would enter into a working vacation period. In this state, if no user arrives when the working vacation period is completed, the channels will enter into the next working vacation period. However, if some newly arriving users enter the channels when the system is in a working vacation period, the users will be served with a lower service rate. If there

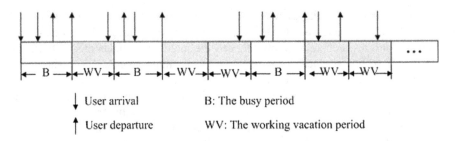

| User arrival | B: The busy period |
| User departure | WV: The working vacation period |

Fig. 1. The dynamic channel vacation scheme proposed in this paper.

are still some users in the channels at the end of the vacation, the vacation period will end and a new busy period will begin.
2. Preemptive priority state. In this state, if there are no channels available but not all channels are used by PUs, the service for SUs will be preempted by a newly arriving PU, and the SU being served will return to the head of the waiting queue.
3. Blocking state. In this state, all channels are used by PUs. The system will discard any newly arriving PUs.

Because of the infinite buffer for the SUs, the SUs are able to access the buffer space directly when waiting for the service if the system is in the second or third state. The specific actions of the PUs and SUs in this paper are as follows. For PUs, if there are some empty channels, a newly arriving PU will be served directly. If there is no channel available but not all channels are being used by PUs, the service of SUs is preempted by newly arriving PUs. If all channels are used by PUs, then any newly arriving PUs will be lost due to the blocked system. For SUs, if there are some empty channels, any newly arriving SU can be served directly. If there are no channels available, any arriving SUs will enter the buffer and wait for the central controller to allot them channels. And if any PUs arrive at the system while SUs are receiving service, the SUs will immediately release their channels to the PUs and enter the buffer at the head of the queue. We assume that the service order is a FIFO discipline.

2.2 System Model

In this model, we assume that SUs are regarded as class I users, PUs are regarded as class II users, and c licensed channels are regarded as c servers. At the same time, considering the preemptive priority of PUs to licensed channels, a continuous time three-dimensional queueing model is built, where the class II user has preemptive priority. In order to ultimately enhance the response performance of PUs, an infinite buffer is set for the SUs, and no buffer is set for the PUs. The inter-arrival time, the service time and the vacation time are all assumed to be independently sequenced. The specific description for this model is as follows:

1. The two classes of users arrive at the system obeying a Poisson process with parameters $\lambda_1, \lambda_2(\lambda_1, \lambda_2 > 0)$ respectively as follows:

$$P\{T_1 \leq x\} = 1 - e^{-\lambda_1 x}, \quad P\{T_2 \leq x\} = 1 - e^{-\lambda_2 x}, x > 0$$

where T_1 and T_2 represent the inter-arrival times of SUs and PUs respectively.

2. The service times S_1, S_2 follow exponential distributions with parameters $\mu_1, \mu_2 (\mu_1, \mu_2 > 0)$ as follows:

$$P\{S_1 \leq x\} = 1 - e^{-\mu_1 x}, \quad P\{S_2 \leq x\} = 1 - e^{-\mu_2 x}, x > 0$$

where S_1, S_2 represent the service times of SUs and PUs when the channels are in busy periods respectively.

3. The vacation time V follows an exponential distribution with parameter θ, and the service times $S_1^{(v)}, S_2^{(v)}$ follow exponential distributions with parameters $\mu_1^{(v)}, \mu_2^{(v)} \left(\mu_1^{(v)}, \mu_2^{(v)} > 0 \right)$ in a working vacation period as follows:

$$P\{V \leq x\} = 1 - e^{-\theta x}, \ P\{S_1^{(v)} \leq x\} = 1 - e^{-\mu_1^{(v)} x}, \ P\{S_2^{(v)} \leq x\} = 1 - e^{-\mu_2^{(v)} x}, x > 0$$

where $S_1^{(v)}, S_2^{(v)}$ represent the service times of SUs and PUs when the channels are in working vacation periods respectively.

3 Model Analysis

According to the above description, let $N_1(t), N_2(t)$ be the number of SUs and PUs in the system at instant t, and $J(t)$ be the server state at instant t. Define

$$J(t) = \begin{cases} 0, & \text{the instant } t \text{ is in the working vacation period,} \\ 1, & \text{the instant } t \text{ is in the busy period.} \end{cases}$$

Then, the three-dimensional stochastic process $\{(N_1(t), N_2(t), J(t)), t \geq 0\}$ is a Markov process with the state space:

$$\Omega = \{(0,\ 0,\ 0)\} \cup \{(0,\ l,\ j),\ 1 \leq l \leq c,\ j = 0,\ 1\} \cup \{(i,\ l,\ j),\ i \geq 1,\ 0 \leq l \leq c,\ j = 0,\ 1\}.$$

All possible states: $(i, 0, 0), (i, 0, 1), (i, 1, 0), (i, 1, 1), \ldots, (i, c, 0), (i, c, 1)$ are called level i, where $i \geq 1$. Specifically, level 0 has states: $(0, 0, 0), (0, 1, 0), (0, 1, 1), \ldots, (0, c, 0), (0, c, 1)$.

Using the lexicographical ordering for the states, the state transition rate matrix of the process can be written as:

$$Q = \begin{bmatrix} A_{00} & A_{01} & & & & & & \\ A_{10} & A_{11} & A_2 & & & & & \\ & A_{21} & A_{22} & A_2 & & & & \\ & & \ddots & \ddots & \ddots & & & \\ & & & A_{c,c-1} & A_{cc} & A_2 & & \\ & & & & A_{c,c-1} & A_{cc} & A_2 & \\ & & & & & \ddots & \ddots & \ddots \end{bmatrix}. \tag{1}$$

According to the block tri-diagonal structure of the transition rate matrix, we can obtain that $\{(N_1(t), N_2(t), J(t)), t \geq 0\}$ a quasi birth and death process. When the Markov process is positive recurrent, the steady-state distribution is indicated as follows.

$$\pi_{i,l,j} = \lim_{t \to \infty} P\{N_1(t)) = i, \ N_2(t) = l, \ J(t) = j\}, \quad (i, \ l, \ j) \in \Omega$$

$$\Pi = (\pi_0, \ \pi_1, \ \pi_2, \ \ldots)$$

where

$$\pi_0 = (\pi_{0,0,0}, \ \pi_{0,1,0}, \ \pi_{0,1,1}, \ \ldots, \ \pi_{0,c,0}, \ \pi_{0,c,1})$$

$$\pi_i = (\pi_{i,0,0}, \ \pi_{i,0,1}, \ \pi_{i,1,0}, \ \pi_{i,1,1}, \ \ldots, \ \pi_{i,c,0}, \ \pi_{i,c,1}), \quad i \geq 1.$$

The necessary and sufficient condition that the Markov process $\{(N_1(t), N_2(t), J(t)), t \geq 0\}$ is positive recurrent is the matrix quadratic equation

$$R^2 A_{c,c-1} + R A_{cc} + A_2 = 0$$

which has a minimal non-negative solution R, a spectral radius SP $(R) < 1$, and a $(2c^2 + 4c + 1)$ dimensional stochastic matrix:

$$B[R] = \begin{bmatrix} A_{00} & A_{01} & & & & & \\ A_{10} & A_{11} & A_2 & & & & \\ & A_{21} & A_{22} & A_2 & & & \\ & & \ddots & \ddots & \ddots & & \\ & & & A_{c-1,c-2} & A_{c-1,c-1} & A_2 & \\ & & & & A_{c,c-1} & R A_{c,c-1} + A_{c,c} \end{bmatrix} \tag{2}$$

and has a left-zero vector. When the Markov process is positive recurrent, its steady-state distribution satisfies:

$$\begin{cases} (\pi_0, \ \pi_1, \cdots, \ \pi_c) B[R] = 0 \\ \sum_{i=0}^{c-1} \pi_i e + \pi_c (I - R)^{-1} e = 1 \\ \pi_i = \pi_c R^{i-c}, \quad i \geq c \end{cases} \tag{3}$$

where e is an appropriate dimensional column vector with all elements being equal to one.

The proof of Eq. (3) can be obtained by using the equilibrium equation $\mathbf{\Pi Q = 0}$ and the matrix-geometric solution method presented in [6]. The main step is given as follows. Firstly, according to Q and $B[R]$, we can prove that $B[R]$ is a $(2c^2 + 4c + 1)$ dimensional square matrix and $B[R]e = \mathbf{0}$. Secondly, because the Markov process satisfies the steady state equation $\mathbf{\Pi Q = 0}$, we can obtain the matrix geometric distribution $\pi_i = \pi_c R^{i-c}$, $i \geq c$. Thirdly, according to the normalization condition, we can get $\sum_{i=0}^{c-1} \pi_i e + \pi_c (I - R)^{-1} e = 1$.

4 Performance Measures

According to the results from Sect. 3, we can obtain the corresponding formulas for the performance measures in terms of data loss rate, the average queue length of SUs and PUs, the throughput of PUs, the channel utility and so on.

1. The average queue length $E(L_1)$ is defined as the number of SUs in the system. The average number $E(L_2)$ is defined as the number of PUs. Then, we have that

$$E(L_1) = \sum_{k=0}^{\infty} kP(L_1 = k) = \sum_{k=1}^{\infty} \sum_{j=0}^{1} \sum_{i=0}^{c} k\pi_{k,i,j},$$

$$E(L_2) = \sum_{i=0}^{c} iP(L_2 = i) = \sum_{i=1}^{c} \sum_{j=0}^{1} \sum_{k=0}^{\infty} i\pi_{k,i,j}.$$

2. The mean busy period $E(B)$ is defined as the mean time that the system state takes to go from busy to empty. The mean busy period $E(B)$ is given by:

$$E(B) = E(L_1)\frac{1}{\mu_1} + E(L_2)\frac{1}{\mu_2}.$$

3. The channel utility P_u is defined as the probability that the channels are occupied in the process of data communication. The channel utility P_u is given by:

$$P_u = \frac{\min\{E(L_1) + E(L_2), c\}}{c}.$$

4. The data loss rate P_d is defined as the probability that PUs have to disappear because of the blocked state. The data loss rate P_d is given by:

$$P_d = \sum_{k=0}^{\infty} \sum_{j=0}^{1} \pi_{k,c,j}.$$

5   Numerical Experiments

In this section, we provide some numerical results to describe the effect of parameters on performance measures. Taking $\lambda_1 = 6, \mu_1 = 3, \mu_1^{(v)} = 0.5, \mu_2^{(v)} = 1, \theta = 3$.

In Fig. 2, taking $\lambda_2 = 10$, we find that the average queue length of the SUs increases with an increase of μ_2. That is mainly because μ_2 increases, the PUs can be served more quickly, which leads to less opportunity for the SUs to occupy the channels, so the average length of the SUs increases too. When μ_2 is unchanged, $E(L_1)$ decreases with an increase in the value c. This is mainly because c increases, so the SUs have more chances to depart. Hence, the average length of the SUs decreases.

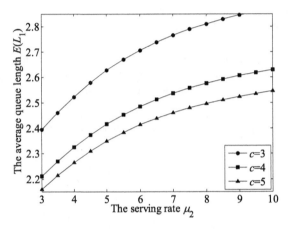

Fig. 2. The relation of $E(L_1)$ with μ_2 and c

In Table 1, taking $\mu_2 = 6$, the average busy period increases with an increase in λ_2. When λ_2 is unchanged, the average busy period decreases with an increase in the value c. This is mainly because an increase of the value c allows more PUs to access service and reduces the opportunity for servers to be in an idle state; therefore, the average busy period decreases.

Table 1. The relation of $E(B)$ with λ_2 and c.

c	The main busy period $E(B)$ of channels				
	$\lambda_2=6$	$\lambda_2=7$	$\lambda_2=8$	$\lambda_2=9$	$\lambda_2=10$
3	0.4900	0.4983	0.5052	0.5109	0.5156
4	0.4678	0.4793	0.4891	0.4975	0.5046
5	0.4594	0.4731	0.4852	0.4957	0.5049

In Fig. 3, taking $\lambda_2 = 10$, we find that the channel utility decreases with an increase in μ_2. This is mainly because an increase in the service rate of the PUs allows users to be serviced quicker, and channels have more chances to be in idle state; therefore the channel utility decreases. When μ_2 is unchanged, the channel utility decreases with an increase in the number of servers. The more channels there are, the more idle channels there will be.

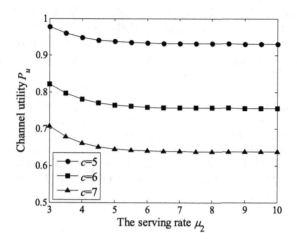

Fig. 3. The relation of P_u with μ_2 and c.

In Fig. 4, taking $\mu_2 = 8$, the data loss rate P_d increases with an increase in λ_2. This is mainly because an increase in the arrival rate of the PUs leads to more channels to be in the blocking state frequently. Hence, the data loss rate P_d increased. On the other hand, when λ_2 is fixed, the data loss rate P_d decreases with an increase in the channel number c. This is because the more channels there are, the more opportunity can be offered to the PUs for service.

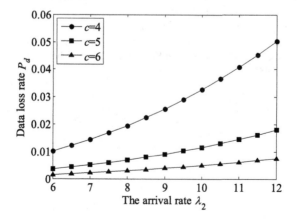

Fig. 4. The relation of P_d with λ_2 and c.

6   Conclusion

In this paper, a dynamic channel vacation scheme in cognitive radio networks is considered in order to conserve network resources and to guarantee the QoS for the users. Based on the working principle of a dynamic channel vacation scheme and a preemptive priority discipline for PUs in cognitive radio networks, we presented a preemptive priority M/M/c queueing model with multiple synchronization working vacations, and established a three-dimension Markov process for the system. By using a matrix-geometric solution method, we derived the steady-state distribution of the queue length. Finally, some numerical examples are discussed to illustrate the effect of the parameters on the system measures.

Acknowledgements. This work was supported in part by the National Natural Science Foundation (No. 61472342), Natural Science Foundation of Hebei Province (No. A2018203088), Key Foundation of Higher Education Science and Technology Research of Hebei Province (No. ZD2017079), Youth Foundation of Higher Education Science and Technology Research of Hebei Province (No. QN2016016), China, and was supported in part by MEXT, Japan, and was supported in part by MEXT and JSPS KAKENHI Grant (Nos. JP17H01825 and JP26280113), Japan.

References

1. Kolodzy, P.: Spectrum policy task force: finding and recommendations. Int. Symp. Adv. Radio Technol. **96**(4), 392–393 (2003)
2. Yu, H., Tang, W., Li, S.: Joint optimal sensing time and power allocation for multi-channel cognitive radio networks considering sensing-channel selection. Sci. China Inf. Sci. **57**(4), 1–8 (2014)
3. Zhao, Y., Jin, S., Yue, W.: Performance optimization of a dynamic channel bonding strategy in cognitive radio networks. Pac. J. Optim. **9**(4), 679–696 (2013)
4. Wu, K., Wang, W., Luo, H., Yu, G., Zhang, Z.: Optimal resource allocation for cognitive radio networks with imperfect spectrum sensing. In: 71st Vehicular Technology Conference (VTC 2010-Spring), pp. 1–4. IEEE (2010)
5. Kim, K.: T-Preemptive priority queue and its application to the analysis of an opportunistic spectrum access in cognitive radio networks. Comput. Oper. Res. **39**(7), 1394–1401 (2012)
6. Neuts, M.: Matrix-Geometric Solutions in Stochastic Models. The Johns Hopkins University Press, Baltimore (1981)

Optimization in Queueing Models

On the Impact of Job Size Variability on Heterogeneity-Aware Load Balancing

Ignace Van Spilbeeck$^{(\boxtimes)}$ and Benny Van Houdt

Department of Mathematics and Computer Science, University of Antwerp - imec,
Middelheimlaan 1, 2020 Antwerp, Belgium
{ignace.vanspilbeeck,benny.vanhoudt}@uantwerpen.be

Abstract. Load balancing is one of the key components in many distributed systems as it heavily impacts performance and resource utilization. We consider a heterogeneous system where each server belongs to one of K classes and the speed of the server depends on its class. Arriving jobs are immediately dispatched to a server class in a randomized manner, i.e., with probability p_k a job is assigned to class k. Within each class a power of d choices rule is used to select the server that executes the job.

For large systems and exponential job size durations the optimal probabilities p_k to minimize the mean response time can be determined easily via convex optimization. In this paper we develop a mean field model (validated by simulation) to investigate how these optimal probabilities p_k are affected by the higher moments and in particular by the variability of the job size distribution when the service discipline at each server is first-come-first-served. The main insight provided is that optimizing the probabilities p_k based on the higher moments is much more involved and provides only a non negligible gain for very specific system load regions.

Keywords: Performance analysis · Distributed computing
Processor scheduling

1 Introduction

Consider a large distributed system consisting of N servers and a (number of) centralized dispatchers. Incoming jobs are assigned by the dispatcher(s) to the servers using a load balancing (LB) scheme. A very efficient manner to distribute the incoming jobs among the servers is to rely on a pure randomized assignment scheme or some form of round robin. While this allows very fast load balancing decisions, the resulting performance is known to be inferior to LB schemes that exploit information concerning the current system state, such as the queue lengths or server speeds. Examples of the latter include join-the-shortest-queue (JSQ) LB [13] or the power-of-d-choices (POD) LB [19,22]. Under JSQ incoming jobs are assigned to the server containing the least number of jobs, while under POD d servers are selected uniformly at random and the job is assigned to the server with the shortest queue length among the d selected servers.

© Springer International Publishing AG, part of Springer Nature 2018
Y. Takahashi et al. (Eds.): QTNA 2018, LNCS 10932, pp. 193–215, 2018.
https://doi.org/10.1007/978-3-319-93736-6_15

When the system is heterogeneous, for instance when not all the servers have the same speed, the choice of the LB scheme becomes even more critical as LB schemes based on joining the server with the least number of jobs among a set of randomly selected servers may lead to system instability even if the total offered load is (well) below the total service rate of the system [6]. A manner to avoid instability when the servers have different speeds (and the overall load is below 1), exists in assigning jobs to servers based on the server speeds [10]. While such an assignment becomes necessary as the overall load tends to one, it is clearly suboptimal under low and medium loads as the mean response time can be reduced by assigning a larger fraction of the jobs to the faster servers. In case of Poisson arrivals, processor sharing (PS) servers and random job routing (that is, server i is selected with a fixed probability p_i) explicit expressions can be derived for the routing probabilities that minimize the mean response time [3,10]. Under first-come-first-served (FCFS) service and more complex LB schemes determining the optimal fraction of the incoming jobs that needs to be assigned to each of the servers is much harder.

In this paper we consider a system consisting of N servers where jobs arrive according to a Poisson process with rate λN with $\lambda < 1$. The servers are partitioned into K classes of homogeneous servers, process their jobs in FCFS order and have an infinite waiting room. By considering FCFS service, we are considering a setting where jobs are very expensive to preempt and are therefore typically run-to-completion without interruption (such as in supercomputing centers, see [14]). Servers belonging to class k serve jobs at rate μ_k and incoming jobs are assigned to a class k server with probability p_k. The server that executes the job within class k is selected using POD LB. In other words, with probability p_k a set of d servers is selected among the class k servers and the jobs is assigned to the server holding the least number of jobs among the d selected class k servers.

Note that the above setting is identical to Scheme 3 presented in [20], except that our servers operate under FCFS instead of PS. For exponential job durations the queue length distribution under FCFS and PS is the same and under PS the system is believed to become insensitive to the job size distribution as the system size N tends to infinity [7,8]. Under FCFS the mean response time remains sensitive to the job size distribution as N tends to infinity. The main objective of this paper is to see how the probabilities p_k that minimize the mean response time in a large system, are effected by the variability of the job size distribution and more importantly whether these optimized values reduce the mean response time significantly compared to the optimal probabilities obtained by assuming exponential job sizes. To answer these questions we develop a mean field model, the accuracy of which is validated using simulation. Our main insights is that neglecting the variability of the job size distribution when optimizing the probabilities p_k does not result in a substantial loss in performance, except under very specific loads combined with highly variable job sizes.

2   The Model

Consider a system of N servers belonging to K classes operating under FCFS. There are N_k servers of class k and let $\gamma_k = N_k/N$ such that $\sum_{k=1}^{K} \gamma_k = 1$. All servers have an infinite waiting room and the speed of a class k server is denoted as μ_k. The server speeds are such that $\sum_{k=1}^{K} \gamma_k \mu_k = 1$, meaning the average speed of a server is equal to 1. Incoming jobs arrive at one or multiple dispatchers as a Poisson process with an overall rate λN and are immediately forwarded to one of the N servers. To select a server the dispatcher selects d servers uniformly at random among the class k servers with a given probability p_k. In other words, the server class is determined via randomization and the server within the class is selected using a POD LB. The job size distribution is assumed to follow a phase-type distribution [17] with mean 1 characterized by (α, S), where α is a stochastic vector and S a subgenerator matrix such that $\alpha e^{Sx} e$ is the probability that the job size exceeds x, where e is a column vector of ones. The time to execute a job on a class k server is therefore phase-type distributed with parameters $(\alpha, \mu_k S)$.

We note that the class of phase-type distributions is dense in the field of all positive-valued distributions. As such any positive-valued distribution can be approximated arbitrarily close by a phase-type distribution. Various fitting tools for phase-type distributions are also available online (e.g., jPhase [21], ProFiDo [5] or BuTools).

Note that due to the Poisson arrivals and randomization, the system under consideration behaves as a set of K independent homogeneous LB systems where the k-th system has load $\rho_k = \lambda p_k/(\gamma_k \mu_k)$ (as the total arrival rate is λN and with probability p_k the job is assigned to one of the γ_k class k servers). For exponential job sizes the probabilities p_k for large N can be optimized by relying on the explicit formula for the mean response time in a homogeneous system derived in [19,22], that is, the probability that a server contains i or more jobs converges to $\rho_k^{\frac{d^i-1}{d-1}}$ as N tends to infinity under POD LB with exponential job sizes and load ρ_k. This results (by Little's law) in the following convex optimization problem that can be solved numerically without much effort:

$$\begin{aligned}
\underset{p_k}{\text{minimize}} \quad & f(p_1, \ldots, p_K) = \frac{1}{\lambda} \sum_k \gamma_k \sum_{i \geq 1} \rho_k^{\frac{d^i-1}{d-1}}. \\
\text{subject to} \quad & 0 \leq \rho_k < 1; k = 1, \ldots, K, \\
& \sum_k \gamma_k \rho_k = \lambda.
\end{aligned} \tag{1}$$

Note that the first set of constraints demands that each of the K subsystems is stable, while the second constraint demands that the total assigned workload matches the incoming workload. For $K = 2$ the first set of constraints can be restated as $1 - \frac{\gamma_2 \mu_2}{\lambda} < p_1 < \frac{\gamma_1 \mu_1}{\lambda}$ (as $p_2 = 1 - p_1$). The main objective of this paper is to study the equivalent optimization problem for phase-type distributed job lengths.

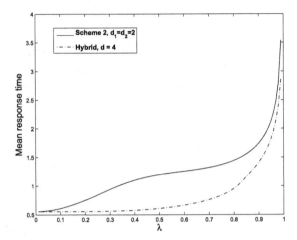

Fig. 1. Mean response time (for exponential job sizes) as a function of λ for Scheme 2 of [1] with $d_1 = d_2 = 2$ and for the hybrid $SQ(4)$ scheme when $\gamma_1 = \gamma_2 = 1/2$ and $\mu_1 = 9\mu_2$.

3   Related Work

A closely related paper is [20] which proposes mean field models for three LB schemes: the optimal randomized, $SQ(d)$ and hybrid $SQ(d)$ LB. The hybrid $SQ(d)$ LB scheme, which was shown to outperform the other two, is identical to the LB scheme considered in this paper except that [20] considers PS servers and exponential job sizes. Evidence that the $SQ(d)$ scheme becomes insensitive to the job size distribution was provided using simulation experiments, while evidence[1] for the asymptotic insensitivity for the hybrid $SQ(d)$ LB under PS was presented in [7,8].

Two other LB schemes for heterogeneous networks were proposed in [1]. In both LB schemes a server is chosen by first selecting d_k servers of type k at random for all k and then by selecting one of the servers among the selected $\sum_k d_k$ servers based on the queue length information only (scheme 1) or on the queue length and server speeds (scheme 2). While Figs. 3 and 4 in [1] suggest that these schemes may outperform the hybrid $SQ(d)$ scheme in some cases, the hybrid $SQ(2)$ scheme uses the queue length information of 2 servers per incoming job, while the other two LB schemes use the queue length information of 4 queues per job. Figure 1 indicates that if we also allow 4 choices for the hybrid $SQ(d)$ scheme, the optimal hybrid $SQ(d)$ scheme outperforms scheme 2 of [1] for all arrival rates λ. Another LB scheme, called HALO_POD, that uses a POD rule in a heterogeneous PS network was proposed in [10]. In this scheme a job is assigned to the shortest of d selected servers, where a class k server is

---

[1] The asymptotic insensitivity under PS was proven given the ansatz of asymptotic independence of the queue length for any finite subset of queues.

selected based on the optimal routing probabilities of a pure randomized LB scheme (first derived in [3]).

In this paper we propose a mean field model to assess the mean response time in a heterogeneous FCFS LB network with phase-type distributed job sizes. Another approach to analyze such a network exists in numerically determining a fixed point of a so-called hydrodynamical PDE presented in [2] as our network is equivalent to a set of K independent homogeneous FCFS networks. In fact, this is the approach that we initially used, but finding the optimal probabilities p_k by repeatedly solving a hydrodynamical PDE turns out to be much more time and memory consuming than the approach taken in this paper (due to the required size of the mesh used by the numerical scheme). Load balancing systems that rely on POD LB with phase-type service were also analyzed in [18], however the set of ODEs presented in [18] is incorrect (as a new job is only assigned to a server if all the d selected servers are in the same service phase).

In this paper we assume that the job sizes are not known in advance. Considerable work has also been performed in case the job sizes are considered to be known when assigning a job to a server. Effective policies in such case include various Size Interval Task Assignment (SITA) and Least Work Left (LWL) policies (e.g., [4,15,16]).

4   The Mean Field Model

Let $X_{k,j,i}^{(N)}(t)$ be the number of type $k \in \{1, \ldots, K\}$ servers with $i > 0$ or more jobs that are in service phase $j \in \{1, \ldots, J\}$ at time t. Define $Z_{k,j,i}^{(N)}(t) = X_{k,j,i}^{(N)}(t)/N_k$ as its scaled version. We would like to study $\lim_{t\to\infty} Z_{k,j,i}^{(N)}(t)$ for large N. For this purpose we introduce a mean field model in Sect. 4.1 for which we provide theoretical and numerical support in Sects. 4.2 and 4.3

4.1   System Dynamics

Assume (α, S) is an order J phase-type distribution. The mean field model uses the variables $s_{k,j,i}(t)$ with $i > 0$, $1 \le j \le J$ and $1 \le k \le K$, that represent the fraction of servers that are of type k, contain i or more jobs and are in service phase j at time t. Let $z_{k,j,i}(t) = s_{k,j,i}(t)/\gamma_k$, denote $\sigma_{i,j}$ as the (i,j)-th entry of the matrix S and let $\nu_j = (-Se)_j$. The evolution of $z_{k,j,i}(t)$ is described by the following set of ODEs, the intuition behind this set of ODEs is presented below:

$$\frac{dz_{k,j,1}(t)}{dt} = \frac{\lambda p_k}{\gamma_k}(1 - z_{k,1}^d(t))\alpha_j - \mu_k \nu_j (z_{k,j,1}(t) - z_{k,j,2}(t))$$

$$- \mu_k \nu_j (1 - \alpha_j) z_{k,j,2}(t) + \sum_{j' \ne j} \mu_k \nu_{j'} z_{k,j',2}(t) \alpha_j$$

$$+ \sum_{j' \ne j} z_{k,j',1}(t) \mu_k \sigma_{j',j} - z_{k,j,1}(t) \sum_{j' \ne j} \mu_k \sigma_{j,j'}$$

$$= \frac{\lambda p_k}{\gamma_k}(1 - z_{k,1}^d(t))\alpha_j - \mu_k z_{k,j,1}(t)\nu_j + \mu_k \sum_{j'=1}^{J} z_{k,j',2}(t)\nu_{j'}\alpha_j$$

$$+ \mu_k \sum_{j'=1}^{J} z_{k,j',1}(t)\sigma_{j',j} - \mu_k z_{k,j,1}(t)\sum_{j'=1}^{J} \sigma_{j,j'}$$

$$= \frac{\lambda p_k}{\gamma_k}(1 - z_{k,1}^d(t))\alpha_j + \mu_k \sum_{j'=1}^{J}(z_{k,j',1}(t)\sigma_{j',j} + z_{k,j',2}(t)\nu_{j'}\alpha_j), \quad (2)$$

where $z_{k,i}(t) = \sum_{j=1}^{J} z_{k,j,i}(t)$ and

$$\frac{dz_{k,j,i}(t)}{dt} = \frac{\lambda p_k}{\gamma_k}\frac{z_{k,j,i-1}(t) - z_{k,j,i}(t)}{z_{k,i-1}(t) - z_{k,i}(t)}(z_{k,i-1}^d(t) - z_{k,i}^d(t))$$

$$- \mu_k\nu_j(z_{k,j,i}(t) - z_{k,j,i+1}(t)) - \mu_k\nu_j(1 - \alpha_j)z_{k,j,i+1}(t)$$

$$+ \sum_{j'\neq j} \mu_k\nu_{j'}z_{k,j',i+1}(t)\alpha_j + \sum_{j'\neq j} z_{k,j',i}(t)\mu_k\sigma_{j',j} - z_{k,j,i}(t)\sum_{j'\neq j}\mu_k\sigma_{j,j'},$$

$$= \frac{\lambda p_k}{\gamma_k}\frac{z_{k,j,i-1}(t) - z_{k,j,i}(t)}{z_{k,i-1}(t) - z_{k,i}(t)}(z_{k,i-1}^d(t) - z_{k,i}^d(t))$$

$$+ \mu_k \sum_{j'=1}^{J}(z_{k,j',i}(t)\sigma_{j',j} + z_{k,j',i+1}(t)\nu_{j'}\alpha_j), \quad (3)$$

for $i > 1$. For $i = 1$ the intuition is as follows. The arrival rate in a class k server is $\lambda p_k/\gamma_k$ and $z_{k,j,1}(t)$ increases when not all of the d selected servers are busy (probability $(1-z_{k,1}^d(t))$) and service starts in phase j (probability α_j). For $i > 1$, $z_{k,j,i}(t)$ increases when all d selected servers have at least $i - 1$ jobs and not all have i jobs or more, this is represented by the probability $(z_{k,i-1}^d(t) - z_{k,i}^d(t))$. The server that gets the job has to be in phase j, which is represented by the probability $\frac{z_{k,j,i-1}(t)-z_{k,j,i}(t)}{z_{k,i-1}(t)-z_{k,i}(t)}$.

For $i \geq 1$, $z_{k,j,i}(t)$ decreases when a job completion occurs in a class k server with exactly i jobs that is in phase j (with rate $\mu_k\nu_j$). It also decreases when a server in phase j with at least $i+1$ jobs has a job completion and starts processing the next job in phase $j' \neq j$ (with rate $\mu_k\nu_j(1 - \alpha_j)$) or a server with at least i jobs changes its phase from j to $j' \neq j$ (with rate $\mu_k\sigma_{j,j'}$). Finally, $z_{k,j,i}(t)$ increases when a server in phase $j' \neq j$ with $i + 1$ or more jobs completes a job and start processing the next job in phase j (with rate $\mu_k\nu_{j'}\alpha_j$) or a server with at least i jobs changes its phase from $j' \neq j$ to j (with rate $\mu_k\sigma_{j',j}$).

Numerical Evaluation: The queue length distribution of the mean field model, characterized by (2–3), is determined via a forward Euler iteration. More specifically, we start with an empty system at time $t = 0$, i.e., set $z_{k,j,i}(0) = 0$ for all k, j and $i > 0$, and compute

$$z_{k,j,i}(t + \delta t) = z_{k,j,i}(t) + \delta t \frac{dz_{k,j,i}(t)}{dt},$$

with a step size δt that is sufficiently small. This iteration is repeated until a fixed point π is found, i.e., until $dz_{k,j,i}(t)/dt \le \epsilon$ for ϵ small (e.g., $\epsilon = 10^{-9}$). The mean response time is subsequently determined via Little's law.

Asymptotic Sensitivity: We end this subsection by showing that any fixed point π of the set of ODEs is sensitive to the higher moments of the job size distribution (as opposed to the system with PS service). Summing (2–3) over i and j yields

$$\sum_{i \ge 1} dz_{k,i}(t)/dt = \mu_k(\rho_k - \sum_{j'} z_{k,j',1}\nu_{j'}).$$

Let ν be the column vector with its j-th entry equal to ν_j and denote β_j as the j-th entry of the unique row vector β for which $\beta(S + \nu\alpha) = 0$ and $\sum_j \beta_j = 1$ holds. It is easy to check that $\beta = \alpha(-S)^{-1}$ and therefore $1/(\beta\nu)$ is the mean job duration. If we now assume asymptotic insensitivity, that is, $\pi_{k,j,i}$ can be written as $\pi_{k,i}\beta_j$, where $\pi_{k,i}$ is the fixed point of the set of ODEs in case of exponential job sizes with load ρ_k, then (3) implies

$$0 = \frac{\lambda p_k}{\gamma_k}\beta_j(\pi_{k,i-1}^d - \pi_{k,i}^d) + \mu_k(\pi_{k,i}(\beta S)_j + \pi_{k,i+1}(\beta\nu)\alpha_j)$$

$$= \frac{\lambda p_k}{\gamma_k}\beta_j(\pi_{k,i-1}^d - \pi_{k,i}^d) - \mu_k(\pi_{k,i} - \pi_{k,i+1})(\beta\nu)\alpha_j, \tag{4}$$

with $\beta\nu = 1$. As $\pi_{k,i}$ is the fixed point of the set of ODEs in case of exponential job sizes with load ρ_k, we have

$$0 = \frac{\lambda p_k}{\gamma_k}(\pi_{k,i-1}^d - \pi_{k,i}^d) - \mu_k(\pi_{k,i} - \pi_{k,i+1}). \tag{5}$$

Hence, (4) holds if and only if $\beta_j = \alpha_j$ for all $j \in \{1, \ldots, J\}$. However, when $\beta = \alpha$ one finds that the probability $\alpha e^{Sx}e$ that the job size exceeds x can be written as

$$\alpha e^{Sx} e = \sum_{s=0}^{\infty} \alpha S^s e x^s/s! = \sum_{s=0}^{\infty}(-\beta\nu)^s x^s/s! = e^{-\beta\nu x},$$

meaning the job sizes are exponential with mean $1/(\beta\nu)$. Thus for any phase-type distribution that is not a redundant[2] representation of the exponential distribution $\pi_{k,i}\beta_j$ is not a fixed point (which would be the case if the system was asymptotically insensitive as in the PS service case).

4.2   Theoretical Support

Let $\mathcal{J} = \{1, \ldots, J\}$ and denote the set of ODEs given by (2–3) as $dz_{k,j,i}(t)/dt = F_{k,j,i}(z_k(t))$, where $z_k(t) = (z_{k,1}(t), z_{k,2}(t), \ldots)$ and $z_{k,i}(t) =$

---

[2] Redundant representations are order J phase-type distributions (α, S) that can be represented by a phase-type distribution of a smaller order. For instance, any order $J > 1$ phase type distribution with S equal to minus the identity matrix is a redundant representation of the exponential distribution with mean one.

$(z_{k,1,i}(t), \ldots, z_{k,J,i}(t))$. Define the space $E^J = \{(x_{j,i})_{j \in \mathcal{J}, i \geq 1} | 1 \geq x_{j,1} \geq x_{j,2} \geq \ldots \geq 0; 1 \geq \sum_{j \in \mathcal{J}} x_{j,1}\}$. Let w be the metric defined on E^J by setting

$$w(\boldsymbol{x}, \boldsymbol{y}) = \sup_{j \in \mathcal{J}} \sup_{i \geq 1} \frac{|x_{j,i} - y_{j,i}|}{(i+1)^2}.$$

Proposition 1. (E^J, w) *is a compact metric space.*

Proof. By Tychonoff's theorem any sequence $(\boldsymbol{x}_n)_n$ in E^J has a subsequence $(\boldsymbol{x}_{n_m})_m$ that converges pointwise to some limit $\boldsymbol{x}^* \in E^J$. We argue that this subsequence also converges to $\boldsymbol{x}^*$ under the metric w which proves compactness. For any i we can pick m' large enough such that for $m \geq m'$ we have

$$\sup_{j \in \mathcal{J}} \frac{|(x_{n_m})_{j,i'} - x^*_{j,i'}|}{(i'+1)^2} \leq 1/(i+1)^2,$$

for $1 \leq i' \leq i$ due to the pointwise convergence. Further

$$\sup_{j \in \mathcal{J}} \frac{|(x_{n_m})_{j,i'} - x^*_{j,i'}|}{(i'+1)^2} \leq 1/(i'+1)^2 < 1/(i+1)^2,$$

for any m when $i < i'$ as $|(x_{n_m})_{j,i'} - x^*_{j,i'}| \leq 1$.   □

The next proposition shows that $\boldsymbol{F}_k(\boldsymbol{x}) : E^J \to E^J$ is Lipschitz, that is, there exists a constant L_k such that $w(\boldsymbol{F}_k(\boldsymbol{x}), \boldsymbol{F}_k(\boldsymbol{y})) \leq L_k w(\boldsymbol{x}, \boldsymbol{y})$. As (E^J, w) is compact it is a Banach space and the Lipschitz property implies that the set of ODEs (2–3) has a unique solution $z_{k,j,i}(t)$ for any given initial state $(z_{k,j,i}(0))_{j \in \mathcal{J}, i \geq 1} \in E^J$ and this solution is continuous in t and the initial state.

Proposition 2. $F_k(\boldsymbol{x})$ *is Lipschitz with constant* $L_k = 3J\mu_k \max_j(-\sigma_{j,j}) + \lambda p_k(2 + dJ + 2Jd^2)/\gamma_k$ *on* (E^J, w).

Proof. We make repeated use of the inequality $|a_1^{m_1} a_2^{m_2} - b_1^{m_1} b_2^{m_2}| \leq m_1|a_1 - b_1| + m_2|a_2 - b_2|$ for $0 \leq a_1, a_2, b_1, b_2 \leq 1$ and $m_1, m_2 \in \{1, 2, \ldots\}$. Due to (2–3) one finds

$$w(\boldsymbol{F}_k(\boldsymbol{x}), \boldsymbol{F}_k(\boldsymbol{y})) \leq 3J\mu_k \max_j(-\sigma_{j,j})w(\boldsymbol{x}, \boldsymbol{y}) + \frac{\lambda p_k}{\gamma_k} dJw(\boldsymbol{x}, \boldsymbol{y}) + \frac{\lambda p_k}{\gamma_k} 2w(\boldsymbol{x}, \boldsymbol{y})$$

$$+ \frac{\lambda p_k}{\gamma_k} \sup_{i > 1} \frac{1}{i+1} \left| \frac{x_{k,i-1}^d - x_{k,i}^d}{x_{k,i-1} - x_{k,i}} - \frac{y_{k,i-1}^d - y_{k,i}^d}{y_{k,i-1} - y_{k,i}} \right|$$

As $(x_{k,i-1}^d - x_{k,i}^d)/(x_{k,i-1} - x_{k,i}) = \sum_{m=0}^{d-1} x_{k,i-1}^m x_{k,i}^{d-1-m}$ we get

$$\sup_{i > 1} \frac{1}{i+1} \left| \frac{x_{k,i-1}^d - x_{k,i}^d}{x_{k,i-1} - x_{k,i}} - \frac{y_{k,i-1}^d - y_{k,i}^d}{y_{k,i-1} - y_{k,i}} \right|$$

$$\leq \sup_{i > 1} \sum_{m=0}^{d-1} \frac{|x_{k,i-1}^m x_{k,i}^{d-1-m} - y_{k,i-1}^m y_{k,i}^{d-1-m}|}{(i+1)^2}$$

$$\leq 2d^2 \sup_{i > 1} \frac{|x_{k,i} - y_{k,i}|}{i+1} \leq 2Jd^2 w(\boldsymbol{x}, \boldsymbol{y}).$$

□

Let $\bar{E}^J = \{(x_{j,i})_{j\in\mathcal{J},i\geq 1} \in E^J \,|\, \sum_{i>0}\sum_{j=1}^J x_{j,i} < \infty\}$, then we have the following result:

Theorem 1. *Let $\boldsymbol{x}(0) \in \bar{E}^J$ and assume $\lim_{N\to\infty} Z_{k,j,i}^{(N)}(0) = x_{j,i}(0)$, then*

$$\lim_{N\to\infty} \sup_{u\leq t} \sup_{j\in\mathcal{J}} \sup_{i\geq 1} \frac{|Z_{k,j,i}^{(N)}(u) - z_{k,j,i}(u)|}{(i+1)^2} = 0 \qquad a.s.,$$

for any fixed t, where $\boldsymbol{z}(u)$ is the unique solution of the set of ODEs given by (2–3) with $z_{k,j,i}(0) = x_{j,i}(0)$.

Proof. The Markov chain $Z_{k,j,i}^{(N)}(t)$, for $N \geq 1$, is a density dependent population process as defined in [9, Chapt. 11]. Theorem 2.1 in [9, Chapter 11] establishes our result provided that two conditions (being (2.6) and (2.7) in [9, Chapt. 11]) apply for any $K \subset \bar{E}^J$ compact. We will argue that both conditions are valid on E^J which implies that they apply to any compact subset of $\bar{E}^J$.

The first condition demands that

$$\sum_{\ell\in L} w(\ell,0) \sup_{\boldsymbol{x}\in E^J} \beta_\ell(\boldsymbol{x}) < \infty,$$

where L is the set of all transitions and $\beta_\ell(\boldsymbol{x})$ is the scaled transition rate of transition ℓ in state $\boldsymbol{x}$. In our system there are three types of transitions (in a queue of length $i > 0$): arrivals, changes in the service phase and service completions. Arrivals in a queue of length i (in service phase j) increase the queue length by one and the vector $\boldsymbol{\ell}$ corresponding to an arrival therefore has two non-zero entries: being $\ell_{j,i}$ which equals -1 and $\ell_{j,i+1}$ which equals $+1$. Hence, $w(\boldsymbol{\ell},0) = 1/(i+1)^2$. Similarly for a change of service phase and a service completion in a queue of length i we find $w(\boldsymbol{\ell},0) = 1/(i+1)^2$.

The scaled rate of any of these transitions for any $\boldsymbol{x} \in E^J$ is bounded by $\lambda p_k/\gamma_k$ (for arrivals) and $\mu_k \max_j(-\sigma_{j,j})$ (for phase changes or service completions). Thus,

$$\sum_{\ell\in L} w(\ell,0) \sup_{\boldsymbol{x}\in E^J} \beta_\ell(\boldsymbol{x}) \leq$$

$$(J\lambda p_k/\gamma_k + J^2\mu_k \max_j(-\sigma_{j,j})) \sum_{i\geq 0} 1/(i+1)^2 < \infty.$$

The second condition demands that $F_k(\boldsymbol{x})$ is Lipschitz, which was shown in Proposition 2. $\qquad\square$

The above theorem indicates that the sample paths of the Markov chains converge to the unique solution of the set of ODEs given by (2–3) as the number of queues N tends to infinity over any finite time scale. One may wonder whether this convergence extends to the stationary regime, meaning whether the steady state measures of the Markov chains weakly converge to the Dirac measure of a fixed point of the set of ODEs. While we believe this to be the case (as indicated in next section that numerically validates this convergence), proving such a result is hard and considered to be out of scope of the current paper.

4.3   Validation

For the model validation we present only results for $K = 2$ types of servers, similar results were obtained for $K > 2$. Let $\mu_r = \frac{\mu_1}{\mu_2}$ and recall that $\gamma_1\mu_1 + \gamma_2\mu_2 = 1$. Further assume that $\mu_1 > \mu_2$, meaning class 1 servers are the *fast* servers and class 2 the *slow* servers. As stated before the mean job size is assumed to be 1. Let C_X^2 be the squared coefficient of variation of the job size distribution. Whenever $C_X^2 = 1/k$ for some $k \in \{2, 3, \ldots\}$, we model the job size distribution as an Erlang distribution with k phases. For $C_X^2 \geq 1$, we used a hyperexponential (HEXP) distribution with parameters (α_1, ν_1, ν_2), thus with probability α_i a job is a type-i job and has an exponential duration with mean $1/\nu_i$, for $i = 1, 2$ (where $\alpha_2 = 1 - \alpha_1$). When $C_X^2 \geq 1$ we additionally match the fraction f of the workload that is contributed by the type-1 jobs (i.e., $f = \alpha_1/\nu_1$). If we assume that $\nu_1 \gg \nu_2$ this can be interpreted as stating that a fraction f of the workload is contributed by the *short* jobs. The mean (equal to 1), C_X^2 and fraction f can be matched as follows:

$$\nu_1 = \frac{C_X^2 + (4f - 1) + \sqrt{(C_X^2 - 1)(C_X^2 - 1 + 8f\bar{f})}}{2f(C_X^2 + 1)}, \tag{6}$$

$$\nu_2 = \frac{C_X^2 + (4\bar{f} - 1) - \sqrt{(C_X^2 - 1)(C_X^2 - 1 + 8f\bar{f})}}{2\bar{f}(C_X^2 + 1)}, \tag{7}$$

with $\bar{f} = 1 - f$ and $\alpha_1 = \nu_1 f$.

Table 1. Parameter settings used to validate the accuracy of the mean field model.

Case	λ	μ_r	γ_1	d	p_1	C_X^2
1	0.26754	1.34	0.6	2	0.1692	0.25
2	0.4116	2.8116	0.4	3	0.79378	0.25
3	0.29374	1.3922	0.7	4	0.47121	0.5
4	0.57975	1.9541	0.5	5	0.53563	0.125
5	0.18995	1.3764	0.3	3	0.43491	0.125
6	0.66992	2.2192	0.6	3	0.71812	0.5
7	0.65294	2.0177	0.4	5	0.57074	4
8	0.24765	1.7567	0.6	2	0.63567	2
9	0.75905	1.6631	0.3	3	0.38569	8
10	0.13251	2.2569	0.5	4	0.22224	4
11	0.78211	2.9592	0.6	5	0.95466	2
12	0.25638	2.8824	0.3	5	0.24434	8

To validate the mean field model, the ODE based mean response times are compared to a discrete event simulation of the system for various parameter settings listed in Table 1. The discrete event simulation has an additional parameter

N which is the size of the system. We let $N \in \{40, 80, 160, 320, 640, 1280\}$ and expect that the mean field model becomes more accurate as N increases. In fact due to the results in [11], the expected response time predicted by the mean field model is $1/N$-accurate, which means that multiplying N by 2 should approximately reduce the relative error by a factor 2. The first six scenarios considered have Erlang distributed job sizes, the last six scenarios have hyperexponentially distributed job sizes where the fraction $f = 1/2$ (for $f \neq 1/2$ similar results were obtained). Table 2 shows the relative error of the mean field model and the associated 95% confidence interval of the simulation runs. In all cases the accuracy improves with N and the relative error is below or close to 10^{-2} for $N \geq 160$. We note that for small N the relative error can be further reduced by relying on the *refined* mean field approximation introduced in [12].

5   Numerical Results

We mainly focus on the case with $K = 2$ types of servers and discuss settings with more than two types of servers in Subsect. 5.4.

5.1   Optimal p_1

In case of exponential job sizes we know that the mean response time is a convex function of p_1 as stated in Sect. 2. Various numerical experiments (see Fig. 2 for one specific example) suggest that the mean response time is still convex in p_1 in case of non-exponential service times. Note that as λ approaches 1, the system is only stable in a very narrow region around $p = \gamma_1 \mu_1$ (which corresponds to a simple proportional assignment). Let p_{opt} be the value of p_1 for which the resulting mean response time is minimized. We now study the impact of the various system parameters on p_{opt}. In Sect. 5.2 we look at the relative increase in the mean response time when a suboptimal p_1 is used.

Arrival Rate λ: As illustrated in Fig. 3 p_{opt} typically decreases as a function of λ (the squares mark the λ value for which the mean response time equals 1). This is expected as fewer jobs in the system implies that one can benefit from sending a larger fraction of the jobs to the fast servers. There are however exceptions, when the job sizes are highly variable and the number of choices is large (e.g., $C_X^2 = 8$ and $d = 20$) the optimal p_1 value may increase as a function of λ at high loads. For λ sufficiently small only the fast servers receive jobs and as $\lambda \to 1$ the load on both server types must be balanced to guarantee stability, i.e., p_1 and p_2 are such that $\frac{\lambda p_1}{\gamma_1 \mu_1} = \frac{\lambda p_2}{\gamma_2 \mu_2}$.

Job Size Variability C_X^2: When looking at the impact of the job size variability C_X^2 in Fig. 3, we note that p_{opt} drops below 1 at lower rates λ when C_X^2 increases. This can be understood by noting that if all the jobs go to the fast servers and λ becomes large enough, some of the jobs start to experience queueing delays. When the job sizes are highly variable, there is a bigger risk of experiencing a

Table 2. Relative error of the mean field model wrt simulation.

Case	$N = 40$ (95% conf.)	$N = 80$ (95% conf.)	$N = 160$ (95% conf.)
1	1.321e−2 (±5.901e−5)	6.439e−3 (±3.939e−5)	3.225e−3 (±3.013e−5)
2	5.484e−3 (±3.682e−5)	2.646e−3 (±2.505e−5)	1.285e−3 (±1.546e−5)
3	2.229e−2 (±5.484e−5)	1.060e−2 (±3.433e−5)	5.277e−3 (±2.656e−5)
4	2.290e−2 (±4.309e−5)	1.078e−2 (±3.063e−5)	5.239e−3 (±2.331e−5)
5	7.873e−4 (±3.726e−5)	3.566e−4 (±2.547e−5)	1.819e−4 (±1.989e−5)
6	3.038e−2 (±6.501e−5)	1.415e−2 (±4.541e−5)	6.720e−3 (±2.900e−5)
7	5.163e−2 (±7.688e−4)	2.260e−2 (±7.466e−4)	1.114e−2 (±3.122e−4)
8	3.935e−3 (±4.969e−4)	1.535e−3 (±3.217e−4)	9.654e−4 (±3.264e−4)
9	9.580e−2 (±2.509e−3)	4.394e−2 (±1.804e−3)	1.937e−2 (±1.028e−3)
10	8.466e−3 (±1.027e−3)	3.486e−3 (±7.692e−4)	2.000e−3 (±4.353e−4)
11	1.631e−1 (±2.260e−3)	7.475e−2 (±1.102e−3)	3.594e−2 (±5.970e−4)
12	1.630e−2 (±1.228e−3)	8.366e−3 (±7.925e−4)	3.983e−3 (±6.873e−4)
Case	$N = 320$ (95% conf.)	$N = 640$ (95% conf.)	$N = 1280$ (95% conf.)
1	1.601e−3 (±1.742e−5)	8.157e−4 (±1.547e−5)	4.197e−4 (±1.081e−5)
2	6.338e−4 (±1.225e−5)	3.003e−4 (±9.128e−6)	1.477e−4 (±7.971e−6)
3	2.709e−3 (±1.710e−5)	1.429e−3 (±1.191e−5)	8.141e−4 (±1.066e−5)
4	2.587e−3 (±1.304e−5)	1.290e−3 (±9.276e−6)	6.288e−4 (±1.115e−5)
5	9.046e−5 (±1.415e−5)	3.764e−5 (±1.043e−5)	1.935e−5 (±5.237e−6)
6	3.109e−3 (±1.994e−5)	1.387e−3 (±1.329e−5)	5.081e−4 (±1.880e−5)
7	5.248e−3 (±2.872e−4)	2.691e−3 (±1.914e−4)	1.332e−3 (±1.746e−4)
8	3.784e−4 (±2.446e−4)	2.082e−4 (±1.665e−4)	1.151e−4 (±1.212e−4)
9	9.372e−3 (±6.956e−4)	5.681e−3 (±3.789e−4)	2.372e−3 (±3.195e−4)
10	9.152e−4 (±4.019e−4)	5.060e−4 (±2.877e−4)	2.400e−4 (±1.696e−4)
11	1.807e−2 (±4.032e−4)	1.012e−2 (±2.764e−4)	6.137e−3 (±2.767e−4)
12	2.280e−3 (±3.967e−4)	1.166e−3 (±2.814e−4)	4.118e−4 (±1.546e−4)

long delay, thus it is advisable to start making use of the slow servers at smaller λ values.

For some parameter settings we see that more variable job sizes result in a lower p_{opt} for any arrival rate λ. This means that when job sizes become more variable, it is beneficial to reduce the fraction of the jobs assigned to the faster servers when minimizing mean response times. This rule is however not valid in all cases: in Fig. 3b, where $d = 5$ and $\mu_r = 2$, we see that p_{opt} for $C_X^2 = 8$ is larger than the corresponding value for $C_X^2 = 4$ for some λ ranges. The cause lies in the fact that the curves of p_{opt} start to oscillate notably for larger d values. These oscillations (that are also visible in Fig. 2b) are probably caused by the fact that for larger d the tail probabilities of the queue length distribution decay very rapidly and depending on the precise value of λ a minor change in λ may

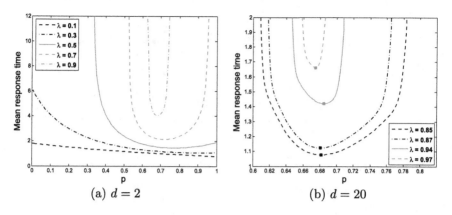

Fig. 2. Mean response time as a function of p for $\gamma_1 = \gamma_2 = 1/2$, $\mu_r = 2$, $f = 1/2$ and $C_X^2 = 4$.

cause a more significant change in the tail probabilities of either the fast or slow servers.

Number of Choices d: Another observation from Fig. 3 is that higher choices for d tend to increase the optimal value of p_1. When d increases the rate λ at which p_1 drops below 1 always seems to increase. This can be understood by noting that increasing d implies that the likeliness of finding an idle fast server when all the jobs are assigned to the fast servers increases. Thus the risk of experiencing a queueing delay decreases with d and therefore assigning all the jobs to the fast servers remains optimal for larger λ values. The fact that p_{opt} increases for increasing d is generally valid for small to medium loads, but does not remain valid for higher loads. For instance it is easily verified that when $\lambda = 0.8$ the p_{opt} for $d = 2$ equals 0.7024, while for $d = 10$ it equals 0.6982 when the job sizes are exponentially distributed.

Higher Moments f: Figure 4 illustrates that the first two moments of the job size distribution do not suffice to determine the optimal split probability p_{opt}, meaning there is no insensitivity with respect to the moments beyond the second moment and optimizing p_{opt} in practice is therefore hard to achieve. The figure also indicates that the optimal fraction of jobs assigned to the fast servers is lower when a larger fraction of the workload consists of *long* jobs. This is intuitively clear: if a larger fraction $1 - f$ of the load is contributed by the long jobs, there is a bigger risk for short jobs to be stuck behind a long job and therefore it is better to make more use of the slow servers.

System Heterogeneity μ_r: We expect that p_{opt} tends to increase as the system heterogeneity $\mu_r = \mu_1/\mu_2$ increases. Figure 5 confirms this intuition for the case with $d = 10$ and 20 choices when $\gamma = f = 1/2$ and $C_X^2 = 1$.

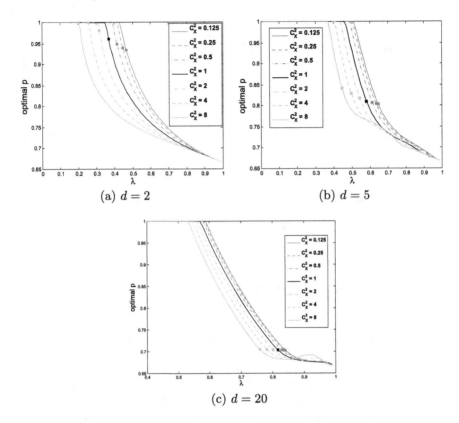

Fig. 3. Optimal choice of p_1 as a function of λ for $\gamma_1 = 0.5$, $\mu_r = 2$, $f = 1/2$ and different values of C_X^2

5.2   Accuracy of Simple Suboptimal Policies

We start by depicting the mean response time for various settings of d, μ_r and C_X^2 in Fig. 6 when using the optimal splitting probability p_{opt}. As expected the mean response time increases with the job size variability, decreases as a function of d and μ_r, and drops below 1 for sufficiently low loads as the mean service time of the fast servers is less than one.

More importantly, one may wonder how much gain in the mean response time one achieves by optimizing p_1. For this purpose we now study the relative gain in the mean response time of the optimal p_1 with the following three less complex assignment policies:

– Proportional: in this case a job is assigned to class k with probability

$$p_k = \frac{\gamma_k \mu_k}{\sum_{j=1}^{K} \gamma_j \mu_j},$$

such that each of the K classes experiences the same load.

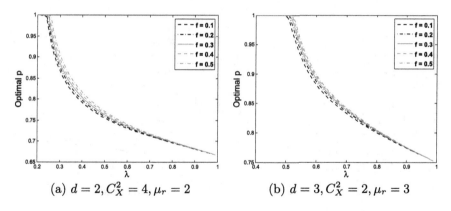

(a) $d = 2, C_X^2 = 4, \mu_r = 2$ (b) $d = 3, C_X^2 = 2, \mu_r = 3$

Fig. 4. Optimal p_{opt} as a function of λ for $\gamma_1 = \gamma_2 = 1/2$.

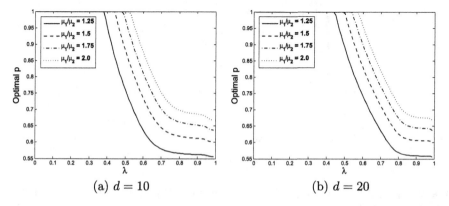

(a) $d = 10$ (b) $d = 20$

Fig. 5. Optimal choice of p_1 as a function of λ for $\gamma_1 = 0.5$, $f = 1/2$, $C_X^2 = 1$ and different values of $\mu_r = \mu_1/\mu_2$.

- Random within a class: in this case we use the optimized probability p_k by assuming that a random server is selected within a class and job lengths are exponential. Hence, p_k is given by the explicit formula in [3,10], which can be written as

$$p_k = \frac{1}{\lambda} \frac{\gamma_k \mu_k}{\sum_{j=1}^{K} \gamma_j \mu_j} + (1 - \frac{1}{\lambda}) \frac{\gamma_k \sqrt{\mu_k}}{\sum_{j=1}^{K} \gamma_j \sqrt{\mu_j}}, \qquad (8)$$

where p_k is set to one (zero) when the above formula results in a p_k larger than one (less than zero). Note as λ approaches one, these probabilities converge to the proportional scheme.
- Exponential job size: in this case we optimize p_1 by solving the convex optimization problem of (1). Hence we optimize assuming that the job lengths are exponential.

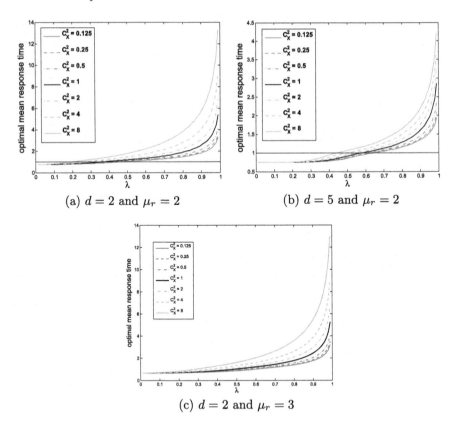

Fig. 6. Optimal mean response times as a function of λ for $\gamma_1 = 0.5$ and $f = 1/2$ for different values of C_X^2.

Note that for all three policies the probabilities $p_1, \ldots, p_K$ only depend on the server speeds and the mean job size (which equals one), either by means of an explicit formula or via a simple convex optimization problem. When $K = 2$ we denote p_1 for the above three policies as p_{prop}, p_{rand} and p_{exp}. Their corresponding mean response times are denoted as $\mathrm{mrt}_{prop}, \mathrm{mrt}_{rand}$ and mrt_{exp}, respectively.

p_{prop} versus p_{opt}: Figure 7 depicts the relative increase in the mean response time if we replace the optimal policy (i.e., p_1 value) with a simple proportional assignment for $\mu_r = 2$. Similar results were obtained for other choices of μ_r. While the proportional scheme is very simple, it results in poor performance for low to medium loads and this loss in performance compared to the optimal policy grows as the number of choices d increases (see Fig. 7b versus c). This is as expected as the optimal strategy under low load exists in sending all the jobs to the fast servers, while the proportional scheme balances the load among the servers.

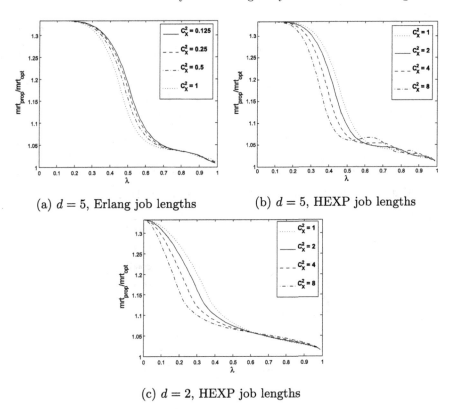

(a) $d = 5$, Erlang job lengths

(b) $d = 5$, HEXP job lengths

(c) $d = 2$, HEXP job lengths

Fig. 7. Ratio $\mathrm{mrt}_{\mathrm{prop}}/\mathrm{mrt}_{\mathrm{opt}}$ as a function of λ with $\mu_r = 2$.

p_{rand} *versus* p_{opt}: Figure 8 depicts the relative increase in the mean response time when relying on (8) instead of using the optimal value for p_k. For small λ both policies (that is, the optimal and the random within a class policy) assign all the jobs to the fast servers. For somewhat higher arrival rates (at about 0.2 in Fig. 8) the random within a class policy starts utilizing the slower servers as well, while the optimal strategy continues to assign all the jobs to the fast servers. Indeed, when all the jobs are assigned to the fast servers, the risk of assigning a job to a busy server increases as d decreases, thus the smaller d the sooner one needs to utilize the slow servers. Figure 8 illustrates that assuming a random assignment (i.e., $d = 1$) results in a performance loss of up to 15% that tends to increase with the number of choices d and that decreases when the job sizes become more variable. The latter can be understood by looking at Fig. 3 which indicates that under low to medium loads, p_{opt} increases as a function of d and decreases as a function of C_X^2. Therefore p_{opt} and p_{rand} are more alike for small d and large C_X^2. We note that in the limit as λ goes to one, both policies use proportional assignment and thus perform alike.

When comparing the relative errors of the proportional scheme with the random within a class policy (compare Figs. 7 and 8), we see that the latter results

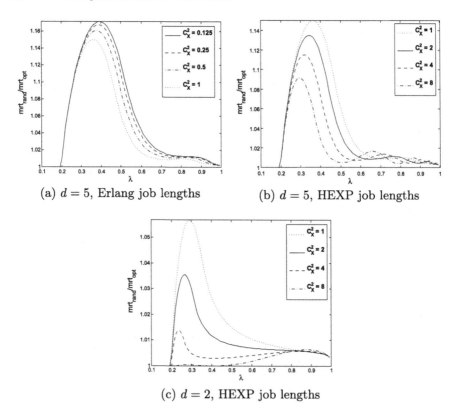

(a) $d = 5$, Erlang job lengths

(b) $d = 5$, HEXP job lengths

(c) $d = 2$, HEXP job lengths

Fig. 8. Ratio $\mathrm{mrt}_{\mathrm{rand}}/\mathrm{mrt}_{\mathrm{opt}}$ as a function of λ with $\mu_r = 2$.

in lower relative errors. We should however note that the proportional scheme is easier to implement as it does require an estimation of the arrival rate λ.

p_{exp} *versus* p_{opt}: Figure 9 studies the relative increase in the mean response time when we only neglect the higher moments of the job size distribution when optimizing p_1. When $d = 5$ this results in errors below 5% and the error is only significant in a fairly small load region. Thus for large enough d, taking the job size variability into account is not paramount (this was confirmed for other μ_r values). When $d = 2$ the relative increase does surge up to 12% in case of highly variable job sizes when λ is close to 0.35. The load at which the relative error is the highest corresponds to the largest arrival rate λ for which p_{exp} still equals 1. Thus, for highly variable job sizes the region where the relative error surges up corresponds to the settings where p_{opt} drops below 1, but p_{exp} remains 1.

Note that solving the convex optimization problem (1) or computing (8) both requires one to estimate the arrival rate λ. When comparing Figs. 8 and 9, it is clear that solving the convex optimization problem (which can be done in a fraction of a second) is far more effective than relying on (8) for $d = 5$, i.e., larger d values. Indeed, the convex optimization problem takes the value

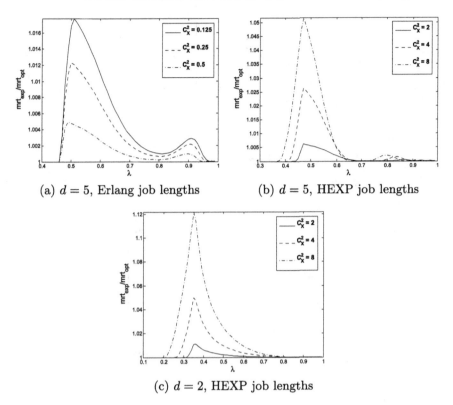

(a) $d = 5$, Erlang job lengths   (b) $d = 5$, HEXP job lengths

(c) $d = 2$, HEXP job lengths

Fig. 9. Ratio $\mathrm{mrt}_{\mathrm{exp}}/\mathrm{mrt}_{\mathrm{opt}}$ as a function of λ with $\mu_r = 2$.

of d into account and therefore causes smaller errors for d large. A somewhat unexpected result is that (8) does result in smaller relative error when $d = 2$, in case of medium loads and highly variable job sizes. The explanation is that when computing p_{rand} we make two errors that mostly cancel each other in this case: we assume that $d = 1$ and that jobs have exponential sizes. For p_{exp} only the latter error is present.

5.3 Impact of the 3rd and Higher Moments of the Job Size Variability

In the previous section we studied the impact of neglecting the job size variability when optimizing p by comparing the performance gain obtained by using p_{opt} instead of p_{exp}. In this section we look at the impact of the higher moments (3rd and beyond). To investigate their impact we consider a hyperexponential distribution as defined in Sect. 4.3, where we matched the mean $EX = 1$, the squared coefficient of variation C_X^2 and the fraction f of the workload contributed by the *short* jobs. Note that changing f influences the higher moments of the job size distribution, but not the mean or variance.

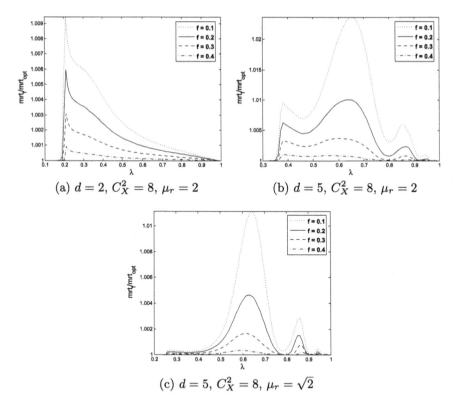

Fig. 10. Ratio mrt$_f$/mrt$_{opt}$ as a function of λ.

To assess the impact of the higher moments we therefore consider job size distributions with $f \neq 1/2$ and compare the mean response time in a system with p_1 optimized for $f = 1/2$, denoted as p_f, with the optimal p_1. Figure 10 depicts the relative gain obtained by using the optimal p_1 instead of p_f when $C_X^2 = 8$ (smaller C_X^2 values result in even smaller relative gains) for $f = 0.1$ to 0.4. While the shape of these curves is very unpredictable and irregular, it is also clear that the relative gain is very minor and less than 2.5% in all cases considered. This indicates that there is little use in taking these higher moments into account when optimizing p_1 (which is good as they are harder to estimate in practice compared to the mean or variance).

5.4 Beyond 2 Server Types

In the previous subsections we assumed that the system consists of two types of servers only. In this section we illustrate that as the number of server types K increases, the differences between the mean response time of the simple policies considered in Subsect. 5.2 and the optimal choice of $p_1, \ldots, p_K$ decreases. In other words, the scenario with $K = 2$ in a way provides an upper bound on how much one gains by optimizing the p_k probabilities.

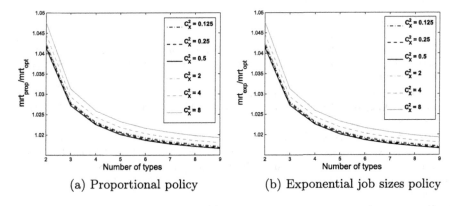

(a) Proportional policy   (b) Exponential job sizes policy

Fig. 11. Relative increase in mean response time when a suboptimal policy is used as a function of the number K of server types for $\lambda = 0.75$, $d = 2$ and $\mu_1/\mu_K = 2$.

In Fig. 11 we set $d = 2$, $\lambda = 0.75$, $\gamma_k = 1/K$ and ordered the server types such that $\mu_1 > \mu_2 > \ldots \mu_K$ with $\mu_1/\mu_K = 2$ and $\mu_{k+1} - \mu_k = \mu_k - \mu_{k-1}$ for $k = 2, \ldots, K - 1$. We depict the increase in the mean job response time when the *proportional* and *exponential job size* policies are used instead of the optimal p_k values with both low and high job size variability (with $f = 1/2$). The results confirm our intuition that this increase tends to diminish as more server types K are introduced.

6   Conclusion

A class of load balancing schemes for a heterogeneous set of FCFS servers is analyzed. The servers are partitioned in K classes of servers: within each class all servers are identical, while servers belonging to different classes only differ in their server speed. Jobs are assigned to a server class via randomization and a power-of-d choices rule is used to select a server within a class. We developed a mean field model to estimate the mean job response time in a system with many servers. The model was supported by some theoretical results and validated using simulation.

While the impact of the different system parameters (like the job size variability or number of choices d) on the optimal randomization probabilities is not always easy to predict (due to oscillations in some of the curves), the main insight provided is that only taking the mean job sizes into account when determining the randomization probabilities (via convex optimization) often results in a very limited loss in performance compared to the optimal probabilities.

The load balancing schemes considered in this paper assume that the job lengths are not known during the job assignment (by the job dispatcher). Further work may exist in studying load balancing schemes that are size aware in combination with a power-of-d choices rule.

References

1. Mukhopadhyay, A.K.A., Mazumdar, R.R.: Randomized assignment of jobs to servers in heterogeneous clusters of shared servers for low delay. Stoch. Syst. **6**(1), 90–131 (2016)
2. Aghajani, R., Li, X., Ramanan, K.: The PDE method for the analysis of randomized load balancing networks. Proc. ACM Meas. Anal. Comput. Syst. **1**(2), 38:1–38:28 (2017)
3. Altman, E., Ayesta, U., Prabhu, B.: Load balancing in processor sharing systems. In: Proceedings of the 3rd International Conference on Performance Evaluation Methodologies and Tools, ValueTools 2008, pp. 12:1–12:10 (2008)
4. Bachmat, E., Sarfati, H.: Analysis of SITA policies. Perform. Eval. **67**(2), 102–120 (2010)
5. Bause, F., Buchholz, P., Kriege, J.: Profido - the processes fitting toolkit dortmund. In: 2010 Seventh International Conference on the Quantitative Evaluation of Systems, pp. 87–96, September 2010
6. Bramson, M.: Stability of join the shortest queue networks. Ann. Appl. Probab. **21**(4), 1568–1625 (2011)
7. Bramson, M., Lu, Y., Prabhakar, B.: Randomized load balancing with general service time distributions. In: ACM SIGMETRICS 2010, pp. 275–286 (2010)
8. Bramson, M., Lu, Y., Prabhakar, B.: Asymptotic independence of queues under randomized load balancing. Queueing Syst. **71**(3), 247–292 (2012)
9. Ethier, S., Kurtz, T.: Markov Processes: Characterization and Convergence. Wiley, New York (1986)
10. Gandhi, A., Zhang, X., Mittal, N.: HALO: heterogeneity-aware load balancing. In: 23rd IEEE MASCOTS 2015, Atlanta, GA, USA, 5–7 October 2015, pp. 242–251 (2015)
11. Gast, N.: Expected values estimated via mean-field approximation are $1/n$-accurate. Proc. ACM Meas. Anal. Comput. Syst. **1**(1), 17:1–17:26 (2017)
12. Gast, N., Van Houdt, B.: A refined mean field approximation. Proc. ACM Meas. Anal. Comput. Syst. **1**(2), 33:1–33:28 (2017)
13. Gupta, V., Harchol Balter, M., Sigman, K., Whitt, W.: Analysis of join-the-shortest-queue routing for web server farms. Perform. Eval. **64**, 1062–1081 (2007)
14. Harchol-Balter, M.: Task assignment with unknown duration. J. ACM **49**(2), 260–288 (2002)
15. Harchol-Balter, M., Crovella, M.E., Murta, C.D.: On choosing a task assignment policy for a distributed server system. J. Parallel Distrib. Comput. **59**(2), 204–228 (1999)
16. Harchol-Balter, M., Scheller-Wolf, A., Young, A.R.: Surprising results on task assignment in server farms with high-variability workloads. SIGMETRICS Perform. Eval. Rev. **37**(1), 287–298 (2009)
17. Latouche, G., Ramaswami, V.: Introduction to Matrix Analytic Methods and Stochastic Modeling. SIAM, Philadelphia (1999)
18. Li, Q.-L., Lui, J.C.S., Wang, Y.: A matrix-analytic solution for randomized load balancing models with PH service times. In: Hummel, K.A., Hlavacs, H., Gansterer, W. (eds.) PERFORM 2010. LNCS, vol. 6821, pp. 240–253. Springer, Heidelberg (2011). https://doi.org/10.1007/978-3-642-25575-5_20
19. Mitzenmacher, M.: The power of two choices in randomized load balancing. IEEE Trans. Parallel Distrib. Syst. **12**, 1094–1104 (2001)

20. Mukhopadhyay, A., Mazumdar, R.R.: Rate-based randomized routing in large heterogeneous processor sharing systems. In: 26th International Teletraffic Congress (ITC), Karlskrona, Sweden, pp. 1–9 (2014)
21. Pérez, J.F., Riaño, G.: jPhase: an object-oriented tool for modeling phase-type distributions. In: Proceeding from the 2006 Workshop on Tools for Solving Structured Markov Chains, SMCtools 2006. ACM, New York (2006)
22. Vvedenskaya, N., Dobrushin, R., Karpelevich, F.: Queueing system with selection of the shortest of two queues: an asymptotic approach. Probl. Peredachi Informatsii **32**, 15–27 (1996)

Analysis of VAS, WAS and XAS Scheduling Algorithms for Fiber-Loop Optical Buffers

Kurt Van Hautegem[1(✉)], Mario Pinto[2], Herwig Bruneel[1], and Wouter Rogiest[1]

[1] SMACS Research Group, Department of Telecommunication and Information Processing (TELIN), Ghent University, Ghent, Belgium
{kurt.vanhautegem,herwig.bruneel,wouter.rogiest}@ugent.be
[2] Universidad Antonio Nariño, Bogota, Colombia
maupinto@uan.edu.co

Abstract. In optical packet/burst switched networks fiber loops provide a viable and compact means of contention resolution. For fixed size packets it is known that a basic void-avoiding schedule (VAS) can vastly outperform a more classical pre-reservation algorithm as FCFS. In this contribution we propose two novel forward-looking algorithms, WAS and XAS, that outperform VAS in the setting of a uniform distributed packet size and a restricted buffer size. This paper presents results obtained by Monte Carlo simulation, showing that improvements of more than 20% in packet loss in specific settings are obtainable. In other settings and for other performance measures similar improvements are within reach.

Keywords: Contention resolution · Scheduling · Optical buffering
Fiber loops

1 Introduction

As video on demand (VoD) services increase in popularity and 4 K video quality will become the new normal, global IP traffic is expected to grow at a compounding annual rate of 24% between 2016 and 2021 [1]. As wavelength and spatial multiplexing allows optical fiber technology to reach dazzling bandwidths of up to 1 Petabit/s [2], it seems that our unlimited demand for bandwidth can be met without a problem. Unfortunately, in existing optical networks capacity is not limited by the connections but in the nodes where slow electronic switching or inflexible optical circuit switching muffle the optical highway capacity.

Promising solutions to address the issues in optical backbones are optical burst switching (OBS) [3–5] and optical packet switching (OPS) [6–8]. In these packet based switching techniques, optical signals are, similar to optical circuit switching (OCS) [9,10], kept in the optical domain to avoid slow optical-electronic-optical (OEO) conversions but, similar to electronic switching, processed as packets to increase statistical multiplexing efficiency. Although RAM

© Springer International Publishing AG, part of Springer Nature 2018
Y. Takahashi et al. (Eds.): QTNA 2018, LNCS 10932, pp. 216–226, 2018.
https://doi.org/10.1007/978-3-319-93736-6_16

buffering in the nodes is infeasible because it requires OEO conversions, at least a limited amount of buffering remains advisable to address the unavoidable contention that arises in the nodes [11,12].

One of the most compact implementations of optical buffering today is a fiber loop buffer. As opposed to feed-forward buffers where every line is traversed only once [13], fiber loop buffers allow contending packets to recirculate multiple times within the same coiled fiber loop [14,15]. Although alternative designs as dual-loop optical buffers exist [16–19], most use a set of single fiber loops in parallel which can all accommodate a single packet at once (shown in Fig. 1). Because packets can only exit a loop after a round number of recirculations, fiber loops can only provide a discrete set of delays. As opposed to electronic memory (RAM) packets can thus not be retrieved at will, resulting in small time gaps or voids in between packets on the outgoing line. Moreover, as packets recirculate in the same loop, fiber loops can only accommodate packet sizes smaller than or equal to their loop length and packet length directly limits the resolution of possible delays. Since the footprint is preferably kept small, with a small number of fiber loops, and also the number of recirculations a packet can make in a loop is kept low to prevent signal degeneration, scheduling algorithms in which the resources are used as efficiently as possible are needed to achieve low packet loss and packet delay.

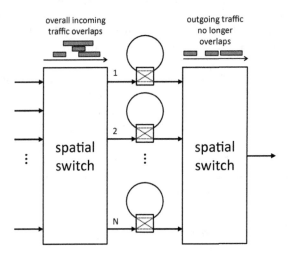

Fig. 1. Parallel optical fiber loop buffer to resolve contention at the input.

In [20] an analytical model is used to evaluate performance of the void-avoiding schedule (VAS) for fixed size packets equal to the loop length. The void-avoiding schedule is a post-reservation scheme [21], allowing the packets to enter the buffer freely, only deciding later when a packet has to exit its loop. In [20] it is shown that performance of the VAS is significantly better than that of algorithms with a pre-reservation scheme, e.g. FCFS, in which the number of recirculations is decided upon arrival of a packet.

In this paper we evaluate the performance of the void-avoiding schedule (VAS) for uniform distributed packet size in both unlimited and constricted fiber loop buffer settings. We also propose two new algorithms, WAS and XAS, which, to the best of our knowledge, are the first known algorithms to outperform VAS. Particularly, both yield significant improvement for variable-length packet size, as discussed further below. The structure of this paper is organized as follows: System model and assumptions are discussed in Sect. 2, the scheduling algorithms in Sect. 3, performance measures and methodology in Sect. 4, numerical results in Sect. 5 and finally conclusions and future work in Sect. 6.

2   System Model and Assumptions

Throughout the paper the same continuous-time setting as in [22–24] is supposed. The fiber loop buffer is assumed to be located at and dedicated to a single outgoing port of an optical switch. Wavelength convertors, if present within the switch, are assumed to perform conversion to a single outgoing wavelength associated with this single outgoing port. The analysis can thus be limited to a single wavelength. We assume the joint packet arrival at the output port on this single wavelength is a Poisson process, i.e. the inter-arrival times T are exponentially distributed with an average of $E[T]$. The length of arriving packets, B, is assumed to be uniform distributed on the interval $[0, S]$ with an average of $E[B] = S/2$. Related, the overall incoming traffic load at the output port is given by $\rho = E[B]/E[T] = S/(2 \cdot E[T])$.

Because of the nature of the arrival process it is possible that different arrivals overlap upon their arrival at which instant one of the contending packets has to be temporarily buffered in one of the fiber loops. We assume a set of parallel fiber loops of length S, that indepently of the packet's size, can accommodate a single packet. The assignable delays to a packet are thus multiples of S. The number of fiber loops and the maximum number of times a packet can recirculate are both varied in simulations. Combinations of both finite (4, 8, and 16) and infinite values for these buffer parameters are evaluated.

3   Scheduling Algorithms

As in a buffer loop setting the well-known FCFS algorithm is outperformed by the void avoiding schedule (VAS), we choose the latter as our benchmark algorithm. In the VAS packets that arrive are transmitted immediately if the outgoing line is available or stored in a fiber loop if not. After each loop recirculation, the availability of the outgoing line is checked. If the outgoing line is available upon such a check, the packet exits its fiber loop and is sent. If the outgoing line is not available, the packet is recirculated again and the procedure is repeated. The VAS does not preserve the arrival order and is a post-reservation algorithm.

In terms of resource usage, VAS is a greedy algorithm, sending a packet whenever a transmission opportunity, i.e. an available outgoing line and either an arrival or a finished recirculation, arises. The newly proposed WAS algorithm, where "W" may refer to the aim of minimizing the **W**ait for the outgoing line to turn available after departure of the two packets, is more considerate and well aware of the other packets present in the system. When a transmission opportunity arises, WAS will first calculate the time needed to send each combination of two packets (packets i and j, $i \neq j$) present in the system. When $N + 1$ packets are present in the system (i.e. packets which are either in the buffer or a new arrival), this gives an $N \times N$ two dimensional matrix with the time needed to send each combination. In this matrix rows are assumed to be the first packet sent and columns the second (rows before columns). Only when the packet is the first packet (i.e. the row, not the column) of the lowest combination in this matrix, the WAS algorithm will send this packet. Otherwise, depending on whether the transmission opportunity is a new arrival or a finished recirculation, this packet is buffered in a new loop or given another round in its loop. Note that in this situation the lowest combination will not necessarily be the next two packets to be sent as the matrix is re-evaluated at every transmission opportunity. Similarly when the packet that triggered the transmission opportunity is actually sent, the packet that was also part of the lowest combination is not guaranteed to be transmitted next.

Similar to the WAS algorithm, the XAS algorithm, where X may refer to the aim of e**X**tending the period during which the outgoing line is effectively used by the two packets, also calculates a combination matrix to decide upon transmission when a transmission opportunity arises. In this matrix the efficiency of the outgoing line is calculated for each combination by dividing the sum of both packet lengths by the total time needed to send each combination. Only when the packet that triggered the transmission opportunity is the first packet of the combination with the highest efficiency, the XAS algorithm will actually send this packet.

When either the number of fiber loops or the maximum number of recirculations is constricted, the WAS and XAS algorithms will transmit a packet upon a transmission opportunity if doing otherwise would result in an immediate and unnecessary loss. Suppose for example that upon arrival of a new packet the output line is available but all of the fiber loops (finite set) are occupied by other packets. In such a case both WAS and XAS will transmit the new arrival even though a more favorable combination may be present in the fiber loops. By doing so the new arrival need not be dropped and unnecessary loss is prevented. Likewise when the maximum number of recirculations is reached upon the end of a loop recirculation, WAS and XAS will always transmit the packet if the output line is available. In the terminology of [25] we could say that both WAS and XAS are tuned to avoid the use of preventive drop. A flowchart showing the decision process of WAS and XAS when both the number of fiber loops and the maximum number of recirculations is constricted is shown in Fig. 2. When either the number of fiber loops and/or the maximum number of recirculations is not constricted, the flowchart slightly simplifies.

Note that in the case of fixed size packets both WAS and XAS schedule in exactly the same way as VAS. Indeed, as all packets have the same size, the combination that minimizes the time to transmit a pair of packets will always consist of the packet causing the transmission opportunity. Only when packet sizes are not equal to a fixed size, WAS and XAS schedule different, and thus possibly better, than VAS.

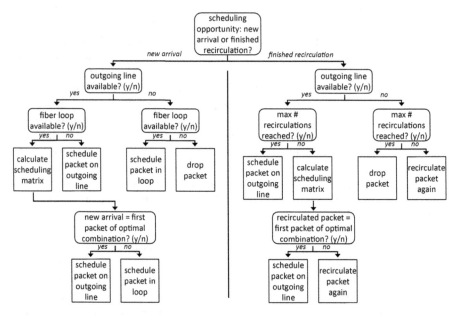

Fig. 2. Flowchart for WAS and XAS both the number of fiber loops and the maximum number of recirculations is constricted.

4   Performance Measures and Methodology

To obtain a complete and representative image of the performance of the various algorithms we look at different performance measures for different settings. For the case with an unlimited number of fiber loops and no restriction on the number of recirculations, no packets are lost and we compare the algorithms on the average packet delay. In case either, or both, the number of fiber loops or the maximum number of recirculations is limited, the loss probability (LP), or equivalently, packet loss, is our main performance measure. In addition, we study a related performance measure which we refer to as LPsize, accounting for the relative total size of packets lost to the relative total size of packets, or, equivalently, the relative amount of data lost with respect to the total amount of data that arrived.

As extending the analytical method from [20] to different algorithms, settings and packet size distributions proved to be too challenging, the performance of the algorithms is evaluated by means of Monte Carlo Simulations. Specifically, all algorithms are programmed in Matlab using a discrete event simulation (DES). In a DES, the system is modelled as a sequence of events marked by their particular instant in time, i.e. the simulation is event-based. The system state changes from one event to the next and does not change in-between events. This is as opposed to continuous simulation in which time is broken into small pieces called time slices. At each ending of a time slice the system state is (possibly) changed based on the events happened in the last time slice. Because DES simulations do not simulate every time slice, they are far more efficient in terms of computational resources.

5   Numerical Results

Fiber loops are assumed to have a length of one time unit and packets to be uniformly distributed on the interval $[0, 1]$. Load is varied from 0.6 to 0.95 in steps of 0.05 by changing the average inter-arrival time of the Poisson arrival process. The arrival of 10^6 packets is simulated 10 times for each algorithm and parameter combination. In this way adequate average performance measures and accompanying confidence intervals are obtained.

Figure 3 shows the average packet delay for the setting with an unlimited number of fiber loops and no restriction on the maximum number of recirculations. As in this setting, for the load values investigated, the FCFS algorithm results in an unstable regime, it is not included in the graph. From Fig. 3 it is clear that in the unrestricted case and with a uniform packet size distribution only XAS can outperform VAS. Table 1 shows the performance improvement (in percentage) XAS can obtain in waiting time relative to VAS. As the load increases, the obtainable improvement also goes up, reaching an improvement of almost 20% for a load of 0.95.

Table 1. Percentage-wise performance improvement in packet delay of XAS relative to VAS for different load values in an unrestricted buffer setting.

Packet delay reduction	Load							
	0.60	0.65	0.70	0.75	0.80	0.85	0.90	0.95
XAS	1.1%	2.4%	4.2%	7.3%	9.6%	13.7%	17.2%	18.1%

Table 2 shows the LP and LPsize of the VAS algorithm in different restricted buffer settings. Note that LP and LPsize have the same values for the VAS algorithm. This is because the length of a packet does not influence the way a packet is scheduled in VAS. In Tables 3 (WAS) and 4 (XAS) the performance improvements (in percentage) compared to VAS of LP and LPsize are shown. This is done for load values of 0.6, 0.7, 0.8 and 0.9, and all combinations of the number of loops and the maximum number of recirculations (both take on values of 4, 8, 16 and infinity). From these tables it is clear that not for all combinations of parameters a performance improvement is possible. In general, but not always, performance improvements increase for lower load values, a lower number of maximum recirculations, and a higher number of loops. Comparing

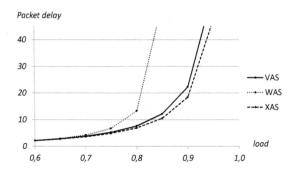

Fig. 3. Packet delay for the VAS, WAS and XAS algorithms in an unrestricted buffer setting.

Table 2. LP and LPsize values for VAS for different load values in different restricted buffer settings.

VAS	Load															
	0.60				0.70				0.80				0.90			
	# loops				# loops				# loops				# loops			
	4	8	16	∞	4	8	16	∞	4	8	16	∞	4	8	16	∞
	$LP = LPsize$ (%)															
Max recirculations 4	10.3	8.8	8.8	8.8	15.7	13.3	13.2	13.2	21.2	18.2	18.0	18.0	26.5	23.0	22.6	22.6
8	7.2	3.6	3.4	3.4	12.6	7.6	6.8	6.8	18.7	12.7	11.2	11.3	24.6	18.4	16.2	16.1
16	6.6	1.5	0.7	0.7	12.0	4.8	2.5	2.4	18.1	10.3	6.1	5.8	24.1	16.7	11.3	10.5
∞	6.6	1.2	0.0	0.0	12.0	4.3	0.4	0.0	18.0	9.8	3.1	0.0	24.1	16.4	9.2	0.0

Table 3. Percentage-wise performance improvement in LP and LPsize of WAS relative to VAS for different load values in different restricted buffer settings.

WAS	Load															
	0.60				0.70				0.80				0.90			
	# loops				# loops				# loops				# loops			
	4	8	16	∞	4	8	16	∞	4	8	16	∞	4	8	16	∞
	LP reduction (%)															
Max recirculations 4	17.3	27.3	27.7	27.4	11.3	21.4	22.0	21.9	7.5	17.7	18.7	18.6	5.2	15.0	16.6	16.5
8	7.0	23.5	23.4	23.7	4.3	16.6	18.2	18.1	2.5	12.6	15.8	15.9	1.3	9.8	15.6	15.7
16	4.1	13.7	−0.7	−1.3	1.7	6.4	−1.6	−3.6	0.5	1.6	2.3	−0.3	−0.6	0.3	6.7	5.5
∞	3.8	21.2	53.6	NaN	1.8	4.8	3.4	NaN	0.2	−4.4	−39.2	NaN	−0.6	−6.9	−32.7	NaN
	LPsize reduction (%)															
Max recirculations 4	8.8	8.8	8.9	8.4	3.4	1.3	1.2	0.8	0.5	−3.2	−3.7	−3.8	−1.2	−5.6	−6.8	−6.8
8	3.4	−3.1	−8.5	−8.1	0.8	−10.1	−16.7	−16.6	−0.8	−11.7	−20.7	−20.6	−1.9	−11.9	−21.5	−21.5
16	3.9	−5.0	−59.8	−61.3	1.2	−10.1	−59.7	−63.4	0.0	−12.2	−51.3	−58.6	−1.2	−11.3	−40.4	−49.5
∞	3.6	21.5	53.0	NaN	1.8	4.9	3.3	NaN	0.3	−4.4	−39.0	NaN	−0.6	−6.8	−32.6	NaN

Table 4. Percentage-wise performance improvement in LP and LPsize of XAS relative to VAS for different load values in different restricted buffer settings.

XAS		Load														
		0.60				0.70				0.80				0.90			
		# loops				# loops				# loops				# loops			
		4	8	16	∞	4	8	16	∞	4	8	16	∞	4	8	16	∞
		LP reduction (%)															
Max recirculations	4	9.9	15.9	16.2	16.3	4.6	8.3	8.7	8.7	1.2	3.2	3.6	3.6	−0.6	−0.1	0.1	0.2
	8	−0.9	10.1	12.4	12.1	−1.0	3.4	5.1	5.5	−1.2	−0.6	−0.2	0.2	−1.6	−2.8	−3.6	−3.4
	16	−3.3	−3.9	0.7	0.7	−2.0	−0.7	1.3	0.7	−1.1	0.2	−2.0	−2.5	−0.8	−0.1	−4.3	−6.2
	∞	−3.8	−4.3	−9.3	NaN	−1.8	2.0	7.6	NaN	−1.1	3.3	13.4	NaN	−0.5	3.2	10.5	NaN
		LPsize reduction (%)															
max recirculations	4	17.5	30.1	30.6	30.6	11.4	22.3	23.2	23.0	7.1	16.2	17.3	17.4	4.6	11.8	13.0	13.2
	8	2.9	31.6	37.9	37.6	2.5	22.0	29.7	30.1	2.1	15.0	22.5	23.0	1.3	10.1	16.9	17.4
	16	−3.1	8.3	41.0	40.9	−1.8	8.5	35.5	36.7	−0.7	7.8	26.3	29.4	−0.4	5.9	18.2	21.8
	∞	−3.8	−4.3	−13.9	NaN	−1.9	2.0	8.0	NaN	−1.1	3.2	13.4	NaN	−0.5	3.3	10.5	NaN

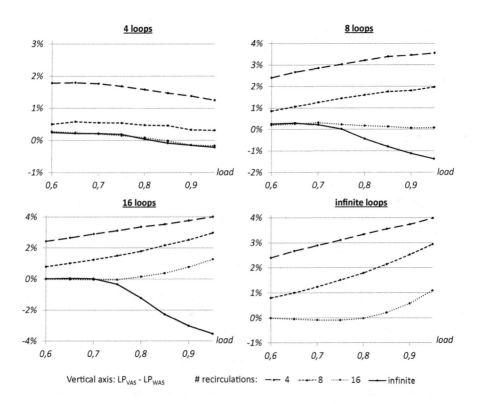

Fig. 4. LP difference between VAS and WAS for different parameter combinations.

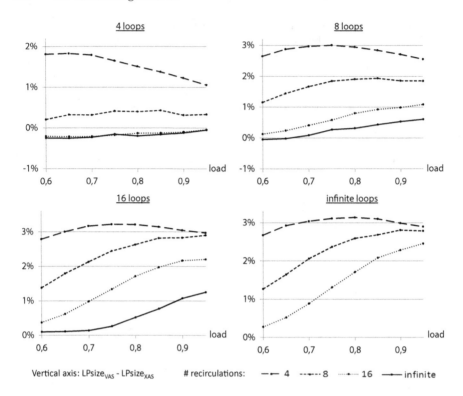

Fig. 5. LPsize difference between VAS and XAS for different parameter combinations.

WAS and XAS, the WAS algorithm seems to be better suited to improve LP with improvements of up to 28% in a setting with a low load (0.6) and a large buffer size (16), but this only for small maximum number of recirculations (4). The XAS algorithm on the other hand seems to be better at improving the LPsize, with improvements of up to 41% for certain parameter combinations, i.e. for a load value of 0.6, a maximum number of recirculations of 16 and a number of loops equal to 16 or infinite.

Note that the trends along certain parameter variations of Tables 3 and 4 are not necessarily visible in the actual performance differences of VAS, WAS and XAS. This is clear from Figs. 4 and 5 showing the difference in LP of VAS and WAS (Fig. 4) and the difference in LPsize of VAS and XAS (Fig. 5) for all parameter combinations. For example although the relative improvement tends to decrease with an increasing load, the actual differences are more steady or even increase. This is as both LP and LPsize increase with an increasing load and a smaller relative improvement can thus correspond with a larger actual performance difference. Similarly, as both LP and LPsize decrease with an increasing number of loops, the relative higher performance improvements for a higher number of loops do not necessarily translate in higher performance differences. This as opposed to the maximum number of recirculations, for which both the relative improvement and the actual performance difference decrease with an increasing number.

6 Conclusions and Future Work

In this paper we proposed the WAS and XAS scheduling algorithms for optical fiber loop buffers in a variable sized packets setting. Performance was evaluated by means of Monte Carlo simulation and showed that they outperform the current state-of-the-art void-avoiding (VAS) schedule in both unlimited and restricted buffer settings. Both algorithms succeed in doing so by "looking ahead", i.e. by taking into account the schedule of other packets present in the system. In this way XAS is capable of improving packet delay with almost 20% for high loads in the unlimited buffer setting. In the restricted buffer setting WAS algorithm is better at improving loss probability (LP) while XAS is better at improving LPsize (related to LP, taking into account packet size). Both algorithms succeed to improve performance with dozens of percentages. To improve the performance of both WAS and XAS even further and for a wider parameter range, future work should focus on taking into account more packets simultaneously or using an optimizable threshold to decide among packet transmission orders.

References

1. Cisco Press Release: The Zettabyte Era: Trends and Analysis (2017). https://www.cisco.com/c/en/us/solutions/collateral/service-provider/visual-networking-index-vni/vni-hyperconnectivity-wp.html
2. Nippon Telegraph and Telephone Corporation: One Petabit per Second Fiber Transmission Over a Record Distance of 200 km (2017). https://www.ntt.co.jp/news2017/1703e/pdf/170323a.pdf
3. Verma, S., Chaskar, H., Ravikanth, R.: Optical burst switching: a viable solution for terabit IP backbone. IEEE Netw. **14**(6), 48–53 (2000)
4. Chen, Y., Qiao, C., Yu, X.: Optical burst switching: a new area in optical networking research. IEEE Netw. **18**(3), 16–23 (2004)
5. Xiong, Y., Vandenhoute, M., Cankaya, H.C.: Control architecture in optical burst-switched WDM networks. IEEE J. Sel. Areas Commun. **18**(10), 1838–1851 (2000)
6. El-Bawab, T.S., Shin, J.-D.: Optical packet switching in core networks: between vision and reality. IEEE Commun. Mag. **40**(9), 60–65 (2002)
7. Szczesniak, I.: Overview of optical packet switching. Theor. Appl. Inform. **21**(3–4), 167–180 (2009)
8. Triki, A., Gravey, A., Gravey, P., Morvan, M.: Long-term CAPEX evolution for slotted optical packet switching in a metropolitan network. In: Proceedings of International Conference on Optical Network Design and Modeling (ONDM), pp. 1–6, May 2017
9. Mukherjee, B.: Architecture, control, and management of optical switching networks. In: Proceedings of Photonics in Switching, pp. 43–44, August 2007

10. Heddeghem, W.V., Lannoo, B., Colle, D., Pickavet, M., Musumeci, F., Pattavina, A., Idzikowski, F.: Power consumption evaluation of circuit-switched versus packet-switched optical backbone networks. In: Proceedings of 2013 IEEE Online Conference on Green Communications, pp. 56–63, October 2013

11. Yao, S., Mukherjee, B., Yoo, S.J.B., Dixit, S.: A unified study of contention-resolution schemes in optical packet-switched networks. J. Lightwave Technol. **21**(3), 672–683 (2003)

12. Yoo, M., Qiao, C., Dixit, S.: The effect of limited fiber delay lines on QoS performance of optical burst switched WDM networks. In: Proceedings of 2000 IEEE International Conference on Communications, vol. 2, pp. 974–979, June 2000

13. Tanemura, T., Soganci, I.M., Oyama, T., Ohyama, T., Mino, S., Williams, K.A., Calabretta, N., Dorren, H.J.S., Nakano, Y.: Large-capacity compact optical buffer based on InP integrated phased-array switch and coiled fiber delay lines. J. Lightwave Technol. **29**(4), 396–402 (2011)

14. Burmeister, E., Blumenthal, D., Bowers, J.: A comparison of optical buffering technologies. Opt. Switch. Network. **5**(1), 10–18 (2008)

15. Langenhorst, R., Eiselt, M., Pieper, W., Grosskopf, G., Ludwig, R., Kuller, L., Dietrich, E., Weber, H.G.: Fiber loop optical buffer. J. Lightwave Technol. **14**(3), 324–335 (1996)

16. Liu, A., Wu, C., Lim, M., Gong, Y., Shum, P.: Optical buffer configuration based on a 3×3 collinear fibre coupler. Electron. Lett. **40**, 1017–1019 (2004)

17. Fu, S., Shum, P., Ngo, N.Q., Wu, C., Li, Y., Chan, C.: An enhanced SOA-based double-loop optical buffer for storage of variable-length packet. J. Lightwave Technol. **26**(4), 425–431 (2008)

18. Tian, C.-Y., Wu, C.-Q., Sun, G.-N., Li, X., Li, Z.-Y.: Quality improvement of the dual-wavelength signals in DLOB via power equalization. Optoelectron. Lett. **4**(5), 361–364 (2008)

19. Wang, Y., Wu, C., Wang, Z., Xin, X.: A new large variable delay optical buffer based on cascaded double loop optical buffers (DLOBs). In: Proceedings of 2009 Conference on Optical Fiber Communication, pp. 1–3, March 2009

20. Rogiest, W., Fiems, D., Dorsman, J.-P.: Analysis of fibre-loop optical buffers with a void-avoiding schedule. In: Proceedings of Valuetools 2014, p. 7, December 2014

21. Rostami, A., Chakraborty, S.S.: On performance of optical buffers with specific number of circulations. IEEE Photon. Technol. Lett. **17**(7), 1570–1572 (2005)

22. Van Hautegem, K., Rogiest, W., Bruneel, H.: Scheduling in optical switching: deploying shared wavelength converters more effectively. In: Proceedings of 2014 IEEE International Conference on Communications (ICC), pp. 3418–3424, June 2014

23. Van Hautegem, K., Rogiest, W., Bruneel, H.: Optical switching for variable size packets: improved void filling through selective void creation. In: Proceedings of 11th International Conference on Queueing Theory and Network Applications (QTNA), pp. 1–8, December 2016

24. Van Hautegem, K., Rogiest, W., Bruneel, H.: Improving performance and energy consumption of shared wavelength converters in OPS/OBS. Opt. Switch. Netw. **17**, 38–51 (2015)

25. Lambert, J., Van Houdt, B., Blondia, C.: Single-wavelength optical buffers: non-equidistant structures and preventive drop mechanisms. In: Proceedings of Networking and Electronic Commerce Research Conference (NAEC), pp. 545–555, October 2005

Maximizing Throughput in Zero-Buffer Tandem Lines with Constrained Flexible Servers

Mohammad H. Yarmand and Douglas G. Down$^{(\boxtimes)}$

Department of Computing and Software, McMaster University, Hamilton, Canada
{yarmanmh,downd}@mcmaster.ca

Abstract. For tandem queues with no buffer spaces and both dedicated and flexible servers, we study how flexible servers should be assigned to maximize the throughput. We focus on systems in which flexible servers are constrained to serve at specific stations. With three stations and one or two constrained flexible servers, we completely characterize the optimal policies and compare throughput improvement with systems in which flexible servers are not constrained. Using numerical experiments, we discuss the impact of constrained flexibility on performance and dimensioning of systems.

Keywords: Server allocation · Zero buffer · Tandem queues
Constrained flexible servers

1 Introduction

Consider a tandem queueing network with $N \geq 2$ stations, $M \geq 1$ dedicated servers, and $F \geq 1$ flexible servers. At any given time, each station can be assigned multiple servers and each server can work only on one job, and a job may have at most one server assigned to it. Assume that the service times of each job at station $i \in \{1, \ldots, N\}$ are independent and identically distributed exponential random variables with rate μ_i, i.e. the service rate of each server at the ith station is only dependent on the station.

We assume that dedicated servers are already assigned to stations. We are interested in determining the dynamic assignment policy for flexible servers that maximizes the long-run average throughput. For simplicity, we assume that the travel times for jobs to progress and also the travel times for flexible servers to move between stations are negligible. We also assume there is an infinite supply of jobs in front of the first station and infinite space for jobs completed at the last station. There are no buffer spaces between stations – blocking occurs after service (manufacturing blocking).

In this paper, we study situations where flexible servers are constrained to operate between two adjacent stations. The motivation for this study is that in practice, there might be situations where moving flexible servers among all stations is not possible (or if possible, it might be costly).

© Springer International Publishing AG, part of Springer Nature 2018
Y. Takahashi et al. (Eds.): QTNA 2018, LNCS 10932, pp. 227–237, 2018.
https://doi.org/10.1007/978-3-319-93736-6_17

The concept of "hand-off" for flexible servers was introduced in [9]. Hand-off happens when a flexible server passes the job it is serving to a dedicated server at the same station. Although it is possible to perform hand-off at any time, we let it occur only in the following two cases. When a station has a busy flexible server and a free dedicated server, the flexible server can pass its job to the dedicated server and become free. When a station has a busy flexible server and a blocked dedicated server, jobs can be swapped between the two servers. In either of the two cases, we say a hand-off has taken place. With the insights gained from [9] (Theorems 1 and 2), our problem is significantly simplified, as we are able to restrict our analysis to policies that perform hand-offs as described above and are non-idling. (Note that is such hand-offs were not allowed, then optimal policies may allow idling, to avoid flexible servers being blocked for extended periods of time.)

Any allocation policy should define appropriate actions when blocking or starvation occurs. In cases where there are multiple blocked or starved servers, policies should prioritize resolving blocking or starvation of the involved servers. We show that for systems with three stations, with balanced service rates, the optimal policy clears blocking from the end of the line to the beginning and avoids starving servers.

The literature on tandem lines with multiple servers and finite buffers is not large, see the discussion in van Vuuren et al. [7]. We have previously studied related problems in the assignment of dedicated and flexible servers in zero-buffer tandem lines. In [8], we studied the problem of how to assign dedicated servers in such settings, and in [9], we examined the problem of how to coordinate dedicated and flexible servers, where the locations of where flexible servers could work was not constrained. Constraining the servers in this manner is the focus of the current paper.

Our work has been motivated by related work in two application domains: hospital bed management and assembly line design. In the hospital bed setting, Bekker et al. [1] discuss a number of issues with respect to how to manage flexible hospital beds. In particular, they develop the insight that full flexibility is beneficial for smaller systems, with less flexibility required for larger systems. Our work complements this, as rather than studying the impact of the number of flexible servers (as in [1]), we explore the scenario when movement of servers is constrained. We imagine that such constrained flexibility may be appropriate in such a setting – it may be problematic to move a flexible bed to all locations in a ward/hospital. Our main conclusion is that there can still be significant benefit, even if flexibility is constrained. For further discussion of the bed management problem, see de Bruin et al. [2], Green [3], and Hall [4]. In terms of assembly line design, an example of a zero-buffer tandem line can be seen in Hu et al. [5], where a subset of fixtures can be reconfigured.

This paper is organized as follows. In Sect. 2 we use Markov Decision Process theory to derive the optimal policy for tandem lines with three stations, a dedicated server at each station, and one or two constrained flexible servers. We further show how to employ the Policy Iteration algorithm to construct the optimal policy.

In Sect. 3 we examine larger systems and discuss the impact of constrained flexibility. Finally, Sect. 4 concludes the paper and discusses future work.

2   Tandem Lines with Three Stations

In this section, we study the following four cases: when there is a flexible server which only moves between the first and second stations; when there is a flexible server which only moves between the second and third stations; when there are two flexible servers, one moving only between the first and second stations and the other one moving only between the second and third stations; and finally when there are two flexible servers that can move among all of the stations (which we call the fully flexible case). We provide a brief comparison between the last two cases. Before looking at these cases, we first discuss the framework for such problems with an arbitrary number of servers and stations. The performance metric that we seek to optimize is the throughput.

We use a Markov Decision Process (MDP) model. For our controlled continuous-time Markov chain (CTMC), let S, A, and A_s represent the state space, the action space, and the action space conditioned on the Markov chain being in state $s \in S$, respectively. We assume that the system employs hand-offs in the following manner. If a station has a busy flexible server and an idle dedicated server, then the busy flexible server can pass its job to an idle dedicated server. Also, when a station has a busy flexible server and a blocked dedicated server, the jobs are swapped between the servers. We also assume that servers are non-idling. Theorems 1 and 2 from [9] can be applied to show that an optimal policy performs hand-offs and is non-idling. With these assumptions, the choice of a state is simplified, as we do not need to keep track of which servers (dedicated or flexible) are busy at each station. Thus our choice of state $s \in S$ is defined by the tuple

$$s = (x_1, y_1, \ldots, x_i, y_i, \ldots, x_N)$$

where x_i is the number of busy servers in station i and y_i is the number of blocked servers in station i. Note that we do not need to include y_N in the state, as no servers can be blocked at the last station. We uniformize the CTMC (see Lippman [6]) to convert the continuous time problem to discrete time. The normalization constant that we will employ is denoted by q, and is defined below.

Constructing the transition matrices is a two stage process. One needs to start from the initial state (s) and follow possible actions (a) to determine the new states (s'). The state s' is an intermediate state which does not appear in the transition matrix. The transition between s and s' is immediate. From there, it is possible to follow further transitions and reach new states (s'') with the probabilities defined in the transition matrix. We have:

$$s \xrightarrow{a} s' \xrightarrow{P_a} s''$$

that is reflected in the transition matrix as $P_a(s, s'') = \frac{\gamma_{a,s,s''}}{q}$ where

$$q = \max_{s \in S, a \in A_s} \sum_{s'' \in \{S-s\}} \gamma_{a,s,s''} \text{ and } \gamma_{a,s,s''} \text{ is the transition rate from } s' \text{ to } s'', \text{ and}$$

s' is the state transitioned to from state s using action a.

The size of the action space is $|A| = \binom{F+N-1}{N-1}$. An action $a \in A$ is denoted by $a_{i_1 i_2 \cdots i_K}$, where i_j is the location of the jth flexible server.

2.1 Constrained Servers

We first consider a tandem line with three stations and a dedicated server at each station. Assume a flexible server exists which can only move between the first and second stations. Theorem 1 describes the optimal policy. In the interest of space, we have not included the proof, as it is similar to the proof of Theorem 3. (Theorem 3 is the key analytic result, so as such it is the one result for which we provide a proof outline.)

Theorem 1. *The optimal policy prioritizes clearing blocking. It performs hand-off and allocates the flexible server to the first station, whenever possible. Otherwise the flexible server is assigned to the second station. The optimal policy is independent of the service rates.*

Next, we consider a tandem line with three stations and a dedicated server at each station. Assume a flexible server exists which can only move between the second and third stations. Theorem 2 describes the optimal policy. As for Theorem 1, its proof is not provided.

Theorem 2. *The optimal policy prioritizes clearing blocking. It performs hand-off and allocates the flexible server to the second station, whenever possible. Otherwise the flexible server is assigned to the third station. The optimal policy is independent of the service rates.*

Comparing the policies described in Theorems 1 and 2 with the fully flexible case, described in [9], all policies have similar structures. They clear blocking and send the flexible server to upstream stations whenever possible.

We now move to the case where there are flexible servers between both stations. Unfortunately, the optimal server assignment becomes more complicated. In general, the optimal policy is rate dependent. However, in the case where the service rates are identical across all stations, the optimal policy again clears blocking (here from the end to the beginning). This would suggest that such a policy would be optimal when the service rates are near balanced across stations. Theorem 3 provides the optimal policy for equal service rates in a zero-buffer tandem line with three stations, a dedicated server at each station, and two flexible servers, one constrained between the first and second stations, the second constrained between the second and third stations.

Theorem 3. *Suppose that $\mu_1 = \mu_2 = \mu_3 = \mu$. The optimal policy clears blocking from the end to the beginning (i.e. the policy prioritizes clearing any blocking at the second station over clearing blocking at the first station, when possible).*

Proof. In the proof, we leave μ_1, μ_2, and μ_3 general, as we will later comment on the dependence of the optimal policy on these rates. To be able to construct the discrete-time MDP, the normalization factor is $q = \mu_1 + \mu_2 + \mu_3 + \max\{\mu_1 + \mu_2, \mu_1 + \mu_3, \mu_2 + \mu_3, 2\mu_2\}$. The MDP details are as follows.

$$A = \{a_{12}, a_{13}, a_{22}, a_{23}\}$$

$$
\begin{aligned}
S = \{&(2,0,2,0,1), (2,0,2,0,0), (2,0,1,1,1), (2,0,1,0,2), (2,0,1,0,1), (2,0,1,0,0),\\
&(2,0,0,1,2), (2,0,0,1,1), (2,0,0,0,2), (2,0,0,0,1), (2,0,0,0,0), (1,1,2,0,0), (1,1,1,0,2),\\
&(1,1,1,0,1), (1,1,1,0,0), (1,1,0,1,2), (1,1,0,1,1), (1,0,3,0,1), (1,0,3,0,0), (1,0,2,1,1),\\
&(1,0,2,0,2), (1,0,2,0,1), (1,0,2,0,0), (1,0,1,2,1), (1,0,1,1,2), (1,0,1,1,1), (1,0,0,2,2),\\
&(1,0,0,2,1), (0,1,3,0,1), (0,1,3,0,0), (0,1,2,1,1), (0,1,2,0,2), (0,1,2,0,1), (0,1,2,0,0),\\
&(0,1,1,2,1), (0,1,1,1,2), (0,1,1,1,1), (0,1,1,0,2), (0,1,1,0,1), (0,1,0,2,2), (0,1,0,2,1),\\
&(1,0,1,0,2), (1,0,1,0,1), (1,0,1,0,0), (1,0,0,1,2), (1,0,0,1,1), (1,0,0,0,2), (1,0,0,0,1),\\
&(1,0,0,0,0)\}
\end{aligned}
$$

Let $\bar{s}$ represent the index of a state $s \in S$, according to the order represented above. In what follows, we will use the label $\bar{s}$ interchangeably with the corresponding state s, i.e. $A_1 = A_{(2,0,2,0,1)}$.

$$
A_{\bar{s}} = \begin{cases}
a_{12} & \text{for } \bar{s} = 1, 2\\
a_{13} & \text{for } \bar{s} = 4, 7, 9\\
a_{22} & \text{for } \bar{s} = 18, 19, 29, 30\\
a_{23} & \text{for } \bar{s} = 21, 25, 27, 32, 36, 40\\
\{a_{12}, a_{13}\} & \text{for } \bar{s} = 3, 5, 6, 8, 10, 11\\
\{a_{12}, a_{22}\} & \text{for } \bar{s} = 12\\
\{a_{13}, a_{23}\} & \text{for } \bar{s} = 13, 16, 38, 42, 45, 47\\
\{a_{22}, a_{23}\} & \text{for } \bar{s} = 20, 24, 31, 35\\
\{a_{12}, a_{22}, a_{23}\} & \text{for } \bar{s} = 22, 23, 33, 34, 41\\
\{a_{13}, a_{22}, a_{23}\} & \text{for } \bar{s} = 28\\
\{a_{12}, a_{13}, a_{22}, a_{23}\} & \text{for } \bar{s} = 14, 15, 17, 26, 37, 39, 43, 44, 46, 48, 49
\end{cases}
$$

Our candidate optimal policy d_0 is:

$$
d_0(\bar{s}) = \begin{cases}
a_{12} & \text{for } \bar{s} = 1, 2, 5, 6, 8, 10, 11, 14, 15, 22, 23, 39, 43, 44, 48, 49\\
a_{13} & \text{for } \bar{s} = 3, 4, 7, 8, 9, 17, 26, 28, 37, 42, 45, 46, 47\\
a_{22} & \text{for } \bar{s} = 12, 18, 19, 29, 30, 33, 34\\
a_{23} & \text{for } \bar{s} = 13, 16, 20, 21, 24, 25, 27, 31, 32, 35, 36, 38, 40, 41
\end{cases}
$$

and associated reward function is:

$$
r_{d_0}(\bar{s}) = \begin{cases}
0 & \text{for } \bar{s} = 2, 6, 11, 12, 15, 19, 23, 30, 34, 39, 44, 49\\
\mu_3 & \text{for } \bar{s} = 1, 3, 5, 8, 10, 14, 17, 18, 20, 22, 24, 26, 29, 31, 33, 35, 37, 43, 46, 48\\
2\mu_3 & \text{for } \bar{s} = 4, 7, 9, 13, 16, 21, 25, 27, 28, 32, 36, 38, 40, 41, 42, 45, 47
\end{cases}
$$

Again in the interests of space, we explicitly provide only the first row of the transition matrix $P_{a_{12}}(s, s'')$. It is a straightforward excercise to calculate the remaining rows of $P_{a_{12}}(s, s'')$, as well as the other matrices $P_{a_{i_1 i_2}}(s, s'')$.

$$P_{a_{12}}(1, 18) = \frac{2\mu_1}{q}$$

$$P_{a_{12}}(1, 4) = \frac{2\mu_2}{q}$$

$$P_{a_{12}}(1, 2) = \frac{\mu_3}{q}$$

$$P_{a_{12}}(1, 1) = \frac{q - 2\mu_1 - 2\mu_2 - \mu_3}{q}$$

$$P_{a_{12}}(1, k) = 0, \ k \neq 1, 2, 4, 18.$$

To prove the optimality of d_0, we need to show that $d_1(s) = d_0(s)$, where:

$$d_1(s) = \text{argmax}_{a \in A_s} \{ r(s, a) + \sum_{j \in S} P_a(s, j) h_0(j) \}, \forall s \in S,$$

is the result of one iteration of the Policy Iteration algorithm. This is equivalent to showing that for all s

$$r(s, a) + \sum_{j \in S} P_a(s, j) h_0(j) - \left(r(s, d_0(s)) + \sum_{j \in S} P_a(s, j) h_0(j) \right) \leq 0. \quad (1)$$

When $\mu = \mu_1 = \mu_2 = \mu_3$, we directly verified inequality that (1) holds for each s and therefore $d_0(s)$ is optimal. The algebra is straightforward (but somewhat lengthy). $\square$

If the service rates are arbitrary, there are several states where the optimal action is rate independent. These states are $(2,0,0,0,1)$, $(2,0,0,0,0)$, $(1,0,0,0,1)$, $(1,0,0,0,0)$ with a_{12}; $(1,1,2,0,0)$, $(0,1,2,0,1)$, $(0,1,2,0,0)$ with a_{22}; and $(1,1,1,0,2)$, $(1,1,0,1,2)$, $(0,1,1,2,1)$, $(0,1,1,0,2)$ with a_{23}.

The remaining states have optimal actions which depend on the service rates. For example, for state $(2,0,1,1,1)$, action a_{13} is the optimal action for equal service rates (and also rates sufficiently close to each other). Action a_{12} becomes the optimal choice when the service rate at the first station is much faster than the second station and a constrained flexible server will be required at the second station, but at the same time the third station is fast enough to clear the blocking with its dedicated server. Looking at an extreme set of rates like $\mu_1 = 20, \mu_2 = 1, \mu_3 = 17$ makes it easier to comprehend why clearing blocking is not the immediate chosen action, but this also occurs for less extreme rates. Unfortunately, explicitly characterizing the boundary between the optimality of actions a_{13} and a_{12} appears difficult in general.

Another example is state $(0,1,1,1,1)$, where action a_{13} is the optimal action for equal service rates. Action a_{23} becomes the optimal choice when service rates are skewed such that the first station is much slower than the second and third stations, in which case admitting jobs becomes the priority. An example set of rates where this holds is $\mu_1 = 1, \mu_2 = 4, \mu_3 = 5$.

2.2   Two Fully Flexible Servers

Consider a tandem line with three stations and a dedicated server at each station. Assume two flexible servers exist that can move between all stations. The optimal policy is rate dependent. When the service rates are equal, Theorem 4 describes the optimal policy. Its proof is similar to the proof of Theorem 3.

Theorem 4. *When $\mu_1 = \mu_2 = \mu$, the optimal policy clears blocking from the end to the beginning. The policy sends the flexible servers to the first station, whenever possible.*

Now we compare the structure of the optimal policies with two flexible servers. When the service rates are equal, in both constrained and fully flexible cases, the optimal policies prioritize clearing blocking. Both of the policies coordinate allocations such that flexible servers are freed to send them to the first station. Both of the policies are rate dependent.

In terms of throughput results, Table 1 compares the two policies for a number of workloads. In this table, the entries in the header row represent service rate vectors and the other table entries are throughput values.

Table 1. Comparison of throughputs for different flexibility situations

Flexibility	Rates			
	$(1, 1, 1)$	$(2, 1, 1)$	$(1, 2, 1)$	$(1, 1, 2)$
Dedicated (5)	0.8873	1.2812	1.1186	1.2888
Dedicated (6)	1.3314	1.4779	1.5597	1.4785
Constrained	1.3606	1.6355	1.4974	1.6027
Fully flexible	1.4600	1.6407	1.7252	1.6435
All fully flexible	1.6667	2.0	2.0	2.0

The first row considers an allocation where there is a dedicated server at each station and two of the stations have one extra dedicated server each. For each column of the first row, the highest throughput resulting from different possible server allocations is represented. The second row represents an allocation where there are two dedicated servers at each station. The third and fourth rows give throughput results for systems with a dedicated server at each station, the third row also has two constrained servers as in Theorem 3, the fourth row has two fully flexible servers as in Theorem 4. The final row has all five servers fully flexible.

Examining Table 1 in more detail, we see that moving to a system with a single constrained server between each pair of servers achieves a significant percentage of the gains made by making servers flexible (compare rows one, three and four). So, without increasing the total number of servers, significant performance gains are possible, even with constrained flexibility. Also, we see that with two constrained servers, we essentially have the same throughput with five servers as with six dedicated servers, a potential savings in required resources.

3   Larger Systems

In this section, we examine how our insights for $N = 3$ stations extends to systems with both a larger number of stations and servers. The numerical results in this section are obtained from simulation, where each simulation is a long run (100 million departures).

We begin with systems where there is one constrained flexible server between each pair of stations. The service rates at all stations are equal to one. In Table 2, the first column, N, shows the number of stations. In the case of dedicated servers (represented in the second column), there are two dedicated servers at each station; in the case of constrained flexible servers (represented in the third column), each station has a dedicated server and there are $N - 1$ constrained flexible servers, one between each consecutive pair of stations (so there is one less server in total than for the dedicated system); in the case of fully flexible servers (represented in the fourth column), there are $N - 1$ fully flexible servers and each station has one dedicated server. For the fully flexible server system (here and throughout this section), we use "Policy I" as described in [9]: "clear blocking from end to beginning only if it does not cause starvation in the $\lfloor \frac{2}{3}N \rfloor$ previous stations; uses hand-off", which was shown to perform well for longer lines. Looking at Table 2, constrained flexibility offers roughly 40% of the throughput improvement that full flexibility provides. Note that the constrained flexibility appears to make the resulting throughput relatively insensitive to N. As a result, the throughput gain (as a percentage) appears to be a slightly increasing function of N.

Table 2. Throughput for larger homogeneous systems

N	Dedicated	Const flx	Fully flx
4	1.2420	1.3438	1.6001
5	1.1939	1.3434	1.6520
8	1.1192	1.3514	1.7286
15	1.0610	1.3561	1.7868
30	1.0282	1.3585	1.8213

We extend our study to systems where there is more than one dedicated server per station. Table 3 considers a configuration with three stations. The third column of this table shows the number of constrained flexible servers among each pair of stations. For example (2,2,0) means that there are two constrained flexible servers between the first and second stations and two constrained flexible servers between the second and third stations. Comparing the first and eighth rows, a configuration with 21 dedicated servers and six constrained flexible servers has throughput close to a configuration with 30 dedicated servers, meaning constrained flexibility can compensate for a reduction of three servers. Also, comparing the eighth and ninth rows, it appears that when there are multiple servers per station, the throughput difference between constrained and full flexibility is less compared to configurations which have one or two servers per stations.

Table 3. Throughput for configurations with $N = 3$ and multiple dedicated and flexible servers per station

N	Dedicated Alloc	Const flx Alloc	Throughput
30	(10, 10, 10)		8.31077
30	(9,10,9)	(1,1,0)	8.63140
29	(9,9,9)	(1,1,0)	8.39741
30	(8,9,9)	(2,2,0)	8.87239
30	(9,9,8)	(2,2,0)	8.86033
30	(8,8,8)	(3,3,0)	9.14992
28	(8,8,8)	(2,2,0)	8.35096
27	(7,7,7)	(3,3,0)	8.22025
27	(7,7,7)	$F = 6$ (fully)	8.63805

To give an idea of how the effect scales, we examine a system with 10 stations, see Table 4. We see that with $N = 91$ or 92 servers and 18 of these being (constrained) flexible servers, we can achieve close to the same throughput as with $N = 100$ servers, all dedicated. In general, it appears that we can reduce the number of servers by approximately 10 percent by adding a small amount of constrained flexibility. Finally, the gap between constrained flexibility and full flexibility increases with the number of stations. This is not at all surprising, as full flexibility allows the flexible servers to work at all of the stations, providing more opportunities to leverage them.

Table 4. Throughput for configurations with $N = 10$ and multiple dedicated and flexible servers per station

N	Allocations	Throughput
100	dedicated: (10,10,10,10,10,10,10,10,10,10)	7.65041
99	dedicated: (9,9,9,9,9,9,9,9,9,9) const flx: (1,1,1,1,1,1,1,1,1,0)	8.06876
99	dedicated: (9,9,9,9,9,9,9,9,9,9) fully flx: 9	8.81186
89	dedicated: (8,8,8,8,8,8,8,8,8,8) const flx: (1,1,1,1,1,1,1,1,1,0)	7.19769
89	dedicated: (8,8,8,8,8,8,8,8,8,8) fully flx: 9	7.93998
98	dedicated: (8,8,8,8,8,8,8,8,8,8) const flx: (2,2,2,2,2,2,2,2,2,0)	8.30885
98	dedicated: (8,8,8,8,8,8,8,8,8,8) fully flx: 18	9.26543
88	dedicated: (7,7,7,7,7,7,7,7,7,7) const flx: (2,2,2,2,2,2,2,2,2,0)	7.41644
91	dedicated: (7,7,7,7,7,8,7,8,7,8) const flx: (2,2,2,2,2,2,2,2,2,0)	7.61384
92	dedicated: (7,7,7,8,7,8,7,8,7,8) const flx: (2,2,2,2,2,2,2,2,2,0)	7.69727
88	dedicated: (7,7,7,7,7,7,7,7,7,7) fully flx: 18	8.28586
97	dedicated: (7,7,7,7,7,7,7,7,7,7) const flx: (3,3,3,3,3,3,3,3,3,0)	8.37838
97	dedicated: (7,7,7,7,7,7,7,7,7,7) fully flx: 27	9.36535
87	dedicated: (6,6,6,6,6,6,6,6,6,6) const flx: (3,3,3,3,3,3,3,3,3,0)	7.47718
87	dedicated: (6,6,6,6,6,6,6,6,6,6) fully flx: 27	8.38774
96	dedicated: (6,6,6,6,6,6,6,6,6,6) const flx: (4,4,4,4,4,4,4,4,4,0)	8.33007
96	dedicated: (6,6,6,6,6,6,6,6,6,6) fully flx: 36	9.45340
86	dedicated: (5,5,5,5,5,5,5,5,5,5) const flx: (4,4,4,4,4,4,4,4,4,0)	7.39672
86	dedicated: (5,5,5,5,5,5,5,5,5,5) fully flx: 36	8.46073
95	dedicated: (5,5,5,5,5,5,5,5,5,5) const flx: (5,5,5,5,5,5,5,5,5,0)	8.14238
95	dedicated: (5,5,5,5,5,5,5,5,5,5) fully flx: 45	9.36825
85	dedicated: (4,4,4,4,4,4,4,4,4,4) const flx: (5,5,5,5,5,5,5,5,5,0)	7.17594
85	dedicated: (4,4,4,4,4,4,4,4,4,4) fully flx: 45	8.40664

4 Conclusion

Based on our observations, optimal policies under constrained flexibility have a similar structure to optimal policies under full flexibility. All of the policies perform hand-off such that the flexible server is freed to send it to upstream stations. They also clear blocking if any exists. Also as expected, the throughput improvement under constrained flexibility is less compared to full flexibility. Unlike full flexibility, the optimal policy under constrained flexibility is not rate independent for arbitrary configurations. The trade-off between the cost of making servers flexible (fully or constrained) and the throughput improvement can be used to decide if flexible servers should be constrained or not. In the future, it would be instructive to explore if structural results could be developed for systems where the service rates are heterogeneous and where the service distributions are not exponential. It would also be instructive to examine more general structures for how servers are constrained. For example, each flexible server could have a "zone" in which they could work – here, the zones are simply pairs of servers, but these zones could be more general in applications.

Acknowledgments. This work was supported by the Discovery Grant program of the Natural Sciences and Engineering Research Council of Canada.

References

1. Bekker, R., Koole, G., Roubos, D.: Flexible bed allocations for hospital wards. Health Care Manag. Sci. **20**, 453–466 (2017)
2. de Bruin, A., Bekker, R., Zanten, L., Koole, G.: Dimensioning clinical wards using the Erlang loss model. Ann. Oper. Res. **178**, 23–43 (2010)
3. Green, L.: Capacity planning and management in hospitals. In: Brandeau, M., Sainfort, F., Pierskella, W. (eds.) Operations Research and Health Care. International Series in Operations Research & Management Science, pp. 15–41. Springer, Boston (2005). https://doi.org/10.1007/1-4020-8066-2_2
4. Hall, R.: Bed assignment and bed management. In: Hall, R. (ed.) Handbook of Healthcare System Scheduling. International Series in Operations Research & Management Science, pp. 177–200. Springer, Boston (2012). https://doi.org/10.1007/978-1-4614-1734-7_8
5. Hu, S., Ko, J., Weyand, L., ElMaraghy, H., Lien, T., Koren, Y., Bley, H., Chryssolouris, G., Nasr, N., Shpitalni, M.: Assembly system design and operations for product variety. CIRP Ann. Manuf. Technol. **60**, 715–733 (2011)
6. Lippman, S.A.: Applying a new device in the optimization of exponential queuing systems. Oper. Res. **23**(4), 687–710 (1975)
7. van Vuuren, M.: Performance analysis of tandem queues with small buffers. IIE Trans. **41**(11), 882–892 (2009)
8. Yarmand, M.H., Down, D.G.: Server allocation for zero buffer tandem queues. Eur. J. Oper. Res. **230**(3), 596–603 (2013)
9. Yarmand, M.H., Down, D.G.: Maximizing throughput in zero-buffer tandem lines with dedicated and flexible servers. IIE Trans. **47**(1), 35–49 (2015)

Applying Reinforcement Learning to Basic Routing Problem

Sigurður Gauti Samúelsson and Esa Hyytiä$^{(\boxtimes)}$

Department of Computer Science, University of Iceland, Reykjavík, Iceland
sgs31@hi.is

Abstract. Routing jobs to parallel servers is a common and important task in today's computer and communication systems. As each routing decision affects the jobs arriving later, determining the (near) optimal decisions is non-trivial. In this paper, we apply reinforcement learning techniques to the job routing problem with heterogeneous servers and a general cost structure. We study the convergence of the reinforcement learning to a near-optimal policy (that we can determine by other means), and compare its performance against heuristic policies such as Join-the-Shortest-Queue (JSQ) and Shortest-Expected-Delay (SED).

Keywords: Job dispatching · Task assignment · Machine learning
Reinforcement learning · Value function · Parallel servers

1 Introduction

Routing jobs to parallel servers has been a long standing problem class for queueing theory. The problem was first studied by Haight already in 1958 [1]. Today, the same problem arises in many new contexts. For example, when routing data traffic in Internet, alternative routes can be modelled as parallel servers. Similarly, in cloud computing, each task needs to be assigned to one of the available servers. In supercomputing, the time scales are longer but the same fundamental question appears. Moreover, the heterogeneity of computing hardware is increasing both in large-scale systems comprising several (thousands of) physical computers, as well as within a single physical device (cf. GPUs vs. CPUs, and new heterogeneous multi-core architectures for mobile devices).

In this paper, we study an elementary routing (or dispatching) problem to heterogeneous parallel servers subject to a large class of cost structures. Both job inter-arrival times and service times are assumed to be exponentially distributed. The state information is the number of jobs in each server. One of the most popular routing policies is Join-the-Shortest-Queue (JSQ), which chooses the server with the fewest jobs. JSQ has been shown to be optimal in some specific cases, but, especially when the service rates are unequal, the exact analysis of the system becomes surprisingly tedious.

The optimization problem for the optimal routing falls in the category of Markov decision processes (MDPs). However, our state space is countably infinite

© Springer International Publishing AG, part of Springer Nature 2018
Y. Takahashi et al. (Eds.): QTNA 2018, LNCS 10932, pp. 238–249, 2018.
https://doi.org/10.1007/978-3-319-93736-6_18

and optimal routing decisions are difficult to determine. We apply reinforcement learning techniques to this problem [2]. The infinite state space remains as a problem as it is impossible to visit every state (preferably multiple times) in any finite time, and therefore learning the optimal action for every state is impossible.

We work around this by focusing on a finite subset of states where decisions presumably matter the most, and rely on an appropriately chosen heuristic routing elsewhere. Effectively, similarly as in [3], we aggregate states so that the resulting optimization problem has a finite set of states, and then apply the reinforcement learning in this state space. If a good heuristic policy is sufficient, then the first policy iteration step (FPI) can be considered. In this case, it is often possible to determine the corresponding value function analytically given the basic policy is static and the system decomposes [4,5]. The value function can also be estimated by a set of short Monte Carlo simulations at each decision point [6]. In this case, the basic policy can be dynamic.

The main contributions of this paper are as follows: First, we show that the heuristic partitioning of the original infinite state space into two classes yields a computationally efficient optimization problem for which machine learning techniques can be applied. Second, we experiment with different learning parameters to gain insight on how fast a near-optimal policy can be learned. This is important especially when the system parameters evolve in time.

The rest of the paper is organized as follows. The routing problem is formally defined in Sect. 2, to which the reinforcement learning technique is in Sect. 3. Section 4 gives some numerical examples, and Sect. 5 concludes the paper.

2   Model

The model for a parallel server system, illustrated in Fig. 1(a), is as follows:

1. Jobs arrive according to a Poisson process with rate λ.
2. Jobs are routed immediately upon arrival to one of the K servers, where the service time in server i is exponentially distributed with parameter μ_i.
3. We consider the so-called *number-aware* setting, where state $\mathbf{n} = (n_1, \ldots, n_K)$ means that server i has n_i jobs. The state space is thus $\mathcal{X} = \mathbb{N}^K$, where $\mathbb{N}$ denotes the set of natural numbers, $\mathbb{N} = \{0, 1, 2, \ldots\}$.
4. Each state $\mathbf{n}$ has an associated *cost rate* $r_{\mathbf{n}}$ at which the system incurs costs when in state $\mathbf{n}$.

   (a) For the mean response time metric, the cost rate is the number of jobs,

   $$r_{\mathbf{n}} = n_1 + \ldots + n_K.$$

   (b) The costs can also deter the use of some servers by using server-specific weights w_i for response time,

   $$r_{\mathbf{n}} = \sum_i w_i n_i.$$

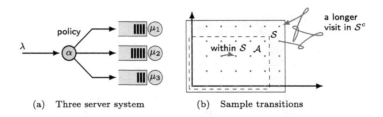

(a)   Three server system   (b)   Sample transitions

Fig. 1. Partitioning the infinite state space by the finite subset S. Visits outside S may involve several jobs arriving and departing before the state of the system returns to S.

(c) If servers incur costs when busy, we have running cost rates,

$$r_\mathbf{n}^{(r)} = \sum_i w_i^{(r)} \mathbf{1}(n_i > 0).$$

Thus, serving a job in server i incurs an average cost of $w_i^{(r)}/\mu_i$. Note that this serves also as an elementary model for energy consumption.

(d) Similarly, with very minor modifications, we can also introduce *admission costs* $c_{i,n}$ incurred when a job enters server i in state n (cf. PASTA).

3   Learning the Optimal Routing Policy

Our aim is to devise a *machine learning* procedure that determines the optimal policy. As mentioned, the state space of the system is infinite, which tends to be a problem as it is not possible to visit every state in finite time. However, often important routing decisions need to be made only in some relatively small subset and elsewhere an appropriate heuristic rule such as Join-the-Shortest-Queue (JSQ) does an adequate job. In particular, decisions near the origin (empty system) are typically critical for the performance.

Therefore, similarly as in [3], we limit our focus on a finite set of states $S \subset \mathcal{X}$. For example, with two server systems we can consider $n \times n$ boxes,

$$S_n = \{(i,j) \mid i < n, \, j < n\}.$$

However, unlike in [3], we are not limited to some specific shapes but S can be arbitrary finite subset of $\mathcal{X}$. The idea is that we will determine the so-called value function only for states in S. Consequently, S induces the action set $\mathcal{A}$ that includes those states $\mathbf{n}$ for which all routing decisions lead to a state in S,

$$\mathcal{A} = \{\mathbf{n} : \mathbf{n} + \mathbf{e}_i \in S \; \forall i\},$$

where $\mathbf{e}_i$ is a vector with the ith component one and all other zero. Hence, given the value function in S is known, it defines the corresponding policy in $\mathcal{A}$.

Elsewhere, in $\mathcal{A}^c$, we assume a fixed heuristic rule such as RND (random split), JSQ or SED (shortest expected delay)[1].

The division of the state space, induced by $\mathcal{S}$, is illustrated in Fig. 1(b). Note that there are (i) direct transitions within $\mathcal{S}$, as well as, (ii) longer visits outside $\mathcal{S}$. For example, long busy periods with many jobs correspond to long visits outside $\mathcal{S}$. Eventually, after a random time T, a stable system still returns to $\mathcal{S}$. The costs C incurred during time T can obviously be high. Nonetheless, both $\mathbb{E}[T]$ and $\mathbb{E}[C]$ can be estimated by straightforward simulations. Effectively, we view states in $\mathcal{S}^c$ as one or more *aggregated super state(s)*: the system "escapes" from $\mathcal{S}$ to somewhere in $\mathcal{S}^c$, and then returns after a random time.

3.1   Learning the Value Function

Let $v(\mathbf{n})$ denote the value function with a fixed routing,

$$v(\mathbf{n}) \triangleq \lim_{t \to \infty} \mathbb{E}[V(\mathbf{n}, t) - rt],$$

where $V(\mathbf{n}, t)$ denotes the cost incurred during time $(0, t)$ when initially in state $\mathbf{n}$ and r is the long-run mean cost rate (assumed to be finite). The value function for any state $\mathbf{n} \in \mathcal{S}$ satisfies (cf. Howard's and Bellman's equations [7,8]),

$$v(\mathbf{n}) = c(\mathbf{n}, \mathcal{S}) - t(\mathbf{n}, \mathcal{S}) \cdot r + \sum_{\mathbf{m} \in \mathcal{S}} p_{\mathcal{S}}(\mathbf{n}, \mathbf{m}) \cdot v(\mathbf{m}), \tag{1}$$

where $c(\mathbf{n}, \mathcal{S})$ denotes the average costs incurred since arriving to state $\mathbf{n}$ until the system moves to a (new) state in $\mathcal{S}$ (that can be the same state $\mathbf{n}$ if the system first moves to a state in $\mathcal{S}^c$), $t(\mathbf{n}, \mathcal{S})$ denotes the corresponding mean time interval, and $p_{\mathcal{S}}(\mathbf{n}, \mathbf{m})$ is the probability that the next state (in $\mathcal{S}$) is $\mathbf{m}$. Equation (1) is the basis for the Reinforcement learning algorithm aiming to find the *optimal control in $\mathcal{A}$*.

Suppose first that the routing is fixed $\omega_0(\mathbf{n})$, and the aim is to determine (estimate) the value function in $\mathcal{S}$ corresponding to $\omega_0(\mathbf{n})$. Let $\mathbf{n}_j \in \mathcal{S}$ denote the jth state visited in $\mathcal{S}$, i.e., $\mathbf{n}_j$ is a sequence of states the system visits from which the states outside $\mathcal{S}$ have been omitted. Then the learning equations for the value function are

$$\begin{aligned} C &\leftarrow C + c_j, \\ T &\leftarrow T + t_j, \\ r &\leftarrow C/T, \\ v(\mathbf{n}_j) &\leftarrow (1 - \alpha_j)v(\mathbf{n}_j) + \alpha_j \left[c_j - t_j \cdot r + v(\mathbf{n}_{j+1}) \right], \end{aligned} \tag{2}$$

where c_j is the costs incurred since entering state $\mathbf{n}_j$ until reaching state $\mathbf{n}_{j+1}$, t_j is the corresponding time interval, and α_j is the learning rate at step j. The

---

[1] RND (random) chooses the server independently in random using some probabilities p_k, JSQ chooses the queue with the least number of jobs, and SED the queue with the shortest expected response time, i.e., the admission cost to queue i is $(n_i + 1)/\mu_i$.

first three equations provide an estimate for the mean cost rate r, and the last equation updates the estimate for the value function. Initially, the learning rate can be set a high value, close to one, and then, as time goes by, it is decreased gradually to zero (or a value close to zero). For example, one can use

$$\alpha_j = e^{-\beta j},$$

where $\beta > 0$ is an appropriately chosen constant. If the system parameters keep on changing, as often is the case in practice, then one can use some fixed small value, e.g., $\alpha = 0.1$.

As the constant offset in the value function is irrelevant (for routing decisions), we can fix it, e.g., so that $v(0) = 0$. In this case, whenever empty state $\mathbf{n} = (0, \ldots, 0)$ is updated, we immediately subtract its new value from all states,

$$v(\mathbf{n}) \leftarrow v(\mathbf{n}) - v(0), \qquad \forall\, \mathbf{n}. \tag{3}$$

Equation (2), combined with (3), learns the value function for states $\mathcal{S}$ for a given routing policy.

3.2 Policy Improvement

Given the value function, one policy iteration round can be carried out, yielding a new routing policy that is better than $\omega_0(\mathbf{n})$ (unless $\omega_0(\mathbf{n})$ was already optimal). This is known as the *first policy iteration* (FPI). In our case, when a job arrives in state $\mathbf{n} \in \mathcal{A}$, the improved policy routes the job to server j such that

$$v(\mathbf{n} + \mathbf{e}_j) \leq v(\mathbf{n} + \mathbf{e}_i) \quad \forall i.$$

Possible ties can be resolved, e.g., in random. Letting $v_0(\mathbf{n})$ denote the value function corresponding to $\omega_0(\mathbf{n})$, the improved routing policy is

$$\omega_1(\mathbf{n}) \triangleq \operatorname*{argmin}_j v_0(\mathbf{n} + \mathbf{e}_j).$$

Example 1. Suppose we have $K = 2$ identical servers, $\mu_1 = \mu_2 = \mu$, and arrival rate $\lambda < 2\mu$. The (basic) routing policy is uniform random split routing a job to server 1 with probability of 0.5, and otherwise to server 2. As the routing decision does not depend on the state of the system, the routing policy is static and the value function decomposes,

$$v(\mathbf{n}) = v_1(n_1) + v_2(n2).$$

Suppose further that the cost structure is the response time metric. Then

$$v_i(n) = \frac{n(n+1)}{2(\mu_i - \lambda_i)} - \frac{\lambda\mu}{(\mu - \lambda)^3},$$

where now $\mu_i = 1$ and $\lambda_i = 0.5 \cdot \lambda$. As the constant in the value functions is irrelevant, we can as well choose $v(0, 0) = 0$, yielding

$$v(\mathbf{n}) = \frac{n_1(n_1 + 1)}{2(\mu - \lambda/2)} + \frac{n_2(n_2 + 1)}{2(\mu - \lambda/2)} = \frac{n_1(n_1 + 1) + n_2(n_2 + 1)}{2\mu - \lambda}. \tag{4}$$

For example, with $\mu = 1$ and $\lambda = 1$, the mean cost rate is $r = 2$ and (4) gives

$$
v(\mathbf{n}) = \begin{bmatrix} 0 & 2 & 6 & 12 & \\ 2 & 4 & 8 & 14 & \cdots \\ 6 & 8 & 12 & 18 & \\ 12 & 14 & 18 & 24 & \\ & \vdots & & & \ddots \end{bmatrix} \tag{5}
$$

In policy iteration, one next determines the value function $v_1(\mathbf{n})$ corresponding to $\omega_1(\mathbf{n})$, yielding a new policy $\omega_2(\mathbf{n})$. This is repeated until the mean cost rate no longer improves and an optimal routing policy has been found. In contrast, with reinforcement learning, one updates the routing policy at the same time as the estimates for the (optimal) value function. This leads to the algorithm described in the next section.

3.3   Reinforcement Learning

Several reinforcement learning techniques have been proposed in the literature. For example, in Q-learning the aim is to learn the utility function $Q(s, a)$ for each state s and corresponding action a. With the optimal policy, one always chooses such action a that maximizes the utility. Q-learning is typically applied to models with a finite horizon or a discounting factor. In this case, the dynamic programming Eq. (1) defining the value function also look different. In particular, there is no need to subtract the mean cost rate.

However, our problem formulation has the infinite time-horizon and the mean cost rate r is an integral part of the dynamic programming Eq. (1), leading to update rules (2). Table 1 describes the complete reinforcement learning algorithm based on (2).

Note also that the reinforcement learning involves two basic modes of operation: *exploration* and *exploitation*. Exploration refers to making random decisions which provide information on the value of actions that currently may seem non-optimal. Exploitation, on the other hand, refers to decisions that utilize the available information and choose (typically) the action that appears to be the optimal. Choosing the ratio between exploration and exploitation is an important optimization problem in reinforcement learning. In our algorithm, we have an implicit function exploit(j) that as a function of time decides (in random) whatever to choose the action that appears optimal (exploit) or to choose a random server (explore). Typically, it is important to explore more at start, but as the time goes by, exploitation should become the default action.

Example 2. Let us continue with the previous example. As a basic policy, we now utilize JSQ outside $\mathcal{A}$. Within $\mathcal{A}$, the routing policy is according to the Reinforcement learning rule. As the system has two identical exponential servers, the optimal routing policy is JSQ also within $\mathcal{A}$. That is, once Reinforcement learning algorithm converges, the resulting value function should be such that

$$
v(\mathbf{n} + \mathbf{e}_1) < v(\mathbf{n} + \mathbf{e}_2) \qquad \forall \, \mathbf{n} \in \mathcal{A},
$$

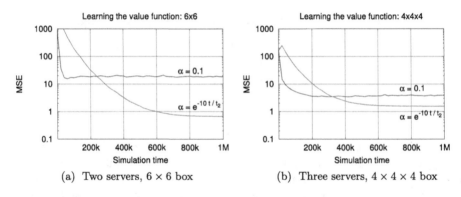

(a) Two servers, 6 × 6 box   (b) Three servers, 4 × 4 × 4 box

Fig. 2. Learning of the value function with a fixed learning rate $\alpha = 0.1$ and when $\alpha = e^{-10t/t_2}$ for two and three server example scenarios. On the x-axis is the time and the y-axis corresponds to the mean squared error (MSE) in log-scale.

whenever $n_1 < n_2$, and vice versa, which means that

$$\omega(\mathbf{n}) = \operatorname*{argmin}_{j} n_j \qquad \forall\, \mathbf{n} \in \mathcal{A},$$

with ties resolved in an arbitrary fashion.

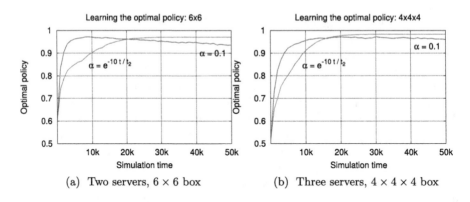

(a) Two servers, 6 × 6 box   (b) Three servers, 4 × 4 × 4 box

Fig. 3. Learning the optimal policy happens much faster.

4   Numerical Examples

In this section, we discuss some numerical experiments with the reinforcement learning. First we assume identical servers so that the correct results are known in advance (see Examples 1 and 2). Then we consider two heterogeneous systems and compare reinforcement learning to some well-known heuristic policies.

Table 1. Reinforcement learning for optimal routing in sub-space $\mathcal{A}$.

Initialization:

$v(\mathbf{x}) \leftarrow 0 \quad \forall \, \mathbf{x} \in \mathcal{S}$

$j \leftarrow 0$ {Step counter}

$\mathbf{x} \leftarrow (0, \ldots, 0)$ {Initial state, $\mathbf{x} \in \mathcal{S}$ }

$\mathbf{n} \leftarrow (0, \ldots, 0)$ {Initial previous state}

$t \leftarrow 0$ {Time between visits in the observed states $\mathcal{S}$}

$c \leftarrow 0$ {Incurred costs between the visits}

$T \leftarrow 0$ {Total (discounted) elapsed time}

$C \leftarrow 0$ {Total (discounted) costs}

After every time step Δt:

$t \leftarrow t + \Delta t \quad$ {Time since the last departure or arrival}

$c \leftarrow c + \Delta c \quad$ {Costs incurred during Δt}

if New job **then**

  if $\mathbf{x} \notin \mathcal{A}$ **then**

    $k \leftarrow \underset{i}{\mathrm{argmin}}\, n_i \quad$ {outside $\mathcal{A}$ use (e.g.) JSQ}

  else if exploit(j) **then**

    $k \leftarrow \underset{i}{\mathrm{argmin}}\, v(\mathbf{n} + \mathbf{e}_i) \quad$ {Ties in random}

  else

    $k \leftarrow \mathrm{random}(1, \ldots, K) \quad$ {Explore, in random}

  end if

  $\mathbf{x} \leftarrow \mathbf{x} + \mathbf{e}_k \quad$ {Send the new job to server k}

else if Departure from server k **then**

  $\mathbf{x} \leftarrow \mathbf{x} - \mathbf{e}_k \quad$ {Remove a job from server k}

end if

if $\mathbf{x} \in \mathcal{S}$ **then**

  $j \leftarrow j + 1$

  $\alpha \leftarrow e^{-\beta j} \quad$ {or α is a small constant}

  $T \leftarrow \gamma T + t$ {e.g., $\gamma = 0.99$}

  $C \leftarrow \gamma C + c$

  $r \leftarrow C/T \quad$ {Mean cost rate}

  $v(\mathbf{n}) \leftarrow (1 - \alpha)v(\mathbf{n}) + \alpha \left[c - t \cdot r + v(\mathbf{x}) \right]$

  if $\mathbf{n} = 0$ **then**

    $\Delta \leftarrow v(0) \quad$ {Adjust offsets}

    for all $\mathbf{n} \in \mathcal{S}$ **do**

      $v(\mathbf{n}) \leftarrow v(\mathbf{n}) - \Delta$

    end for

  end if

  $t \leftarrow 0$ {New epoch starts}

  $c \leftarrow 0$

  $\mathbf{n} \leftarrow \mathbf{x}$

end if

4.1 Learning the Value Function

In the first numerical experiment, we study how fast the value function can be learned. To this end, we assume two or three identical servers with $\mu = 1$, unit arrival rate $\lambda = 1$, and the RND basic policy. The boxes for the substate spaces have 6×6 and $4 \times 4 \times 4$ states, respectively, which (relative) values are to be learned. Moreover, we use either a fixed $\alpha = 0.1$, or let α decay exponentially, $\alpha(t) = e^{-\beta t/t_2}$, where $\beta = 10$ and t_2 is the length of the simulation. The simulation algorithm is otherwise the same as in Table 1, but the server for the new jobs is always chosen using the basic policy, i.e.,

$$k = \omega_0(\mathbf{x}),$$

where ω_0 is RND in our case. That is, we update $v(\mathbf{x})$ but do not use it to make (better) routing decisions. Note that we could learn the value function of any given policy, but we have chosen RND because its value function is known exactly, and we see how fast the system learns it.

Figure 2 depicts the convergence of the learned value function to the known exact solutions, given in (5) for two servers. On the x-axis is the simulation time, and the y-axis corresponds to the mean squared error (MSE),

$$\text{MSE} = \frac{1}{N} \sum_i (\hat{v}_i - v_i)^2,$$

where $N = 6^2$ and $N = 4^3$ in our case. Note that the y-axis is in logarithmic scale. We can see that at with a fixed $\alpha = 0.1$, the learning converges fast to a certain level. When α decreases exponentially (with $\beta = 10$), the learning rate is slower, but the final result is more accurate, as expected. However, the estimates for the value function are useful long before that, as we will see next.

4.2 Learning the Optimal Policy

Next we study how the reinforcement learning algorithm converges to the optimal policy in this elementary case. That is, with the identical servers and the response time cost metric, the optimal policy is JSQ. At start, the value function is initialized to zero, and thus random server would be chosen at every state. However, as different states have been visited and the corresponding updates for the value function recorded, we soon start to make correct decisions.

Figure 3 depicts the fraction of states where the correct routing decision is made as a function of the simulation time. We can observe that the optimal behavior is learned much faster than the "correct" values for the value function. This suggests that a routing policy based on the reinforcement learning can quickly adapt to changes in its operating environment. In such cases, a fixed learning rate such as $\alpha = 0.1$ is naturally preferred.

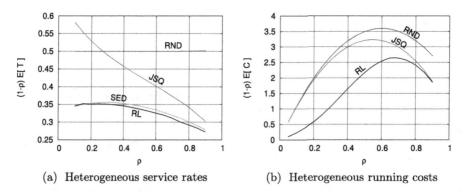

(a) Heterogeneous service rates   (b) Heterogeneous running costs

Fig. 4. Left figure (a) depicts the simulation results with heterogeneous service rates $(\mu_1, \mu_2) = (3, 1)$. Right figure (b) shows the simulation results with unequal running cost rates, $(r_1^{(r)}, r_2^{(r)}) = (0, 10)$.

4.3   Heterogeneous Service Rates

Let us next consider a heterogeneous system with service rates $(\mu_1, \mu_2) = (3, 1)$, i.e., server 1 is now three times faster than server 2. We note that optimal routing policy for heterogeneous systems is not available in closed-form for the mean response time metric even when the service times are exponentially distributed. The near-optimal policy can be determined numerically [3], and here we apply the reinforcement learning algorithm to the same end.

The simulation results are depicted in Fig. 4(a). On the x-axis is the offered load ρ, and the y-axis corresponds to the scaled mean response time, $(1-\rho)\,\mathbb{E}[T]$. With the load balancing random split, the system reduces into K independent $M/M/1$ queues, and the mean response time is

$$\mathbb{E}[T] = \sum_i \frac{\mu_i}{\sum_j \mu_j} \cdot \mathbb{E}[T_i] = \sum_i \frac{1}{\sum_j \mu_j} \frac{1}{1 - \rho} = \frac{K}{(1 - \rho)\sum_j \mu_j},$$

and thus with the two servers the scaled mean response time with RND is $1/2$. Other reference policies are JSQ and SED. The reinforcement learning (RL) uses SED outside the 6×6 box, where the optimal value function is learned and utilized to make near-optimal routing decisions. We can observe that the performance with RL indeed is better than with JSQ or SED.

4.4   Unequal Running Costs

Finally, suppose we have two equally fast servers, $(\mu_1, \mu_2) = (1, 1)$, but server 2 is owned by a third party and they charge us according to used CPU cycles so that the corresponding running costs are $(r_1^{(r)}, r_2^{(r)}) = (0, 10)$. In other words, processing a job at server 2 costs on average $r_2^{(r)}/\mu_2 = 10$, whereas at server 1 it is free. In addition to the running costs, the mean response time is also minimized (i.e., the mean response time is the quality of service component) and the total cost rate at state (n_1, n_2) is given by

$$r_{\mathbf{n}} = n_1 + n_2 + 10 \cdot \mathbf{1}(n_2 > 0).$$

The simulation results are depicted in Fig. 4(b). As $\mu_1 = \mu_2$, SED reduces to JSQ and it has been omitted. On the x-axis is the offered load ρ, and the y-axis corresponds to the scaled mean cost rate, $(1 - \rho) \mathbb{E}[C]$. We can see that all policies, RND, JSQ and RL, have the same shape. Moreover, the dynamic policies, JSQ and RL, seem to converge to the same mean cost rate as $\rho \to 1$. At this limit, both servers must be busy all the time and the unequal running cost rates no longer matter. However, when ρ is small or moderate, the reinforcement learning based policy RL reduces costs significantly. It routes jobs to server 2 only to the extend it is meaningful!

5   Conclusions

A straightforward reinforcement learning approach is studied in this paper. The approach is more general than our numerical examples suggest. First, as mentioned, the cost structure can be rather general and could, e.g., penalize the system when a queue length exceeds given thresholds. Second, without any modifications, the number of servers can be more than two or three. The finite substate space unavoidably becomes larger, which eventually limits the applicability to small systems in terms of number of servers. However, if some servers are identical, the corresponding symmetries can be taken into account to mitigate the scaling problem. Third, it is also straightforward to include *batch arrivals* to the model and the learning algorithm. By adjusting the batch size distribution, more bursty arrival processes can be modelled, which makes the approach more applicable. Fourth, in our case, the jobs were identical. It is possible to introduce job classes, having, e.g., different size distributions or holding cost rates. However, each job class increases the dimensionality of the state space, and therefore we are again limited to a small number of job classes.

In our future work, we plan to investigate on how well the reinforcement learning based dispatching policy adapts to changing environment. In particular, we will compare it to other adaptive and (load) insensitive routing policies.

Acknowledgements. This work was supported by the Academy of Finland in the FQ4BD project (grant no. 296206) and by the University of Iceland Research Fund in the RL-STAR project.

References

1. Haight, F.A.: Two queues in parallel. Biometrika **45**(3–4), 401–410 (1958)
2. Watkins, C.: Learning from delayed rewards, Ph.D. dissertation, Cambridge University (1989)

3. Hyytiä, E., Righter, R., Samúelsson, S.G.: Beyond the shortest queue routing with heterogeneous servers and general cost function. In: ValueTools, December 2017
4. Whittle, P.: Optimal Control: Basics and Beyond. Wiley, New York (1996)
5. Hyytiä, E.: Lookahead actions in dispatching to parallel queues. Perform. Eval. **70**(10), 859–872 (2013). (IFIP Performance 2013)
6. Hyytiä, E., Virtamo, J.: Dynamic routing and wavelength assignment using first policy iteration. In: The Fifth IEEE ISCC 2000, pp. 146–151, July 2000
7. Howard, R.A.: Dynamic Probabilistic Systems, Volume II: Semi-Markov and Decision Processes. Wiley Interscience, New York (1971)
8. Bellman, R.: Dynamic Programming. Princeton University Press, Princeton (1957)

Author Index

Printed in the United States
By Bookmasters